Advances in Silk Research

"This book is dedicated to my beloved mother and late father; to my revered master, my wife and son; and to my beloved readers."

Advances in Silk Research

Dr. N. Gokarneshan

WOODHEAD PUBLISHING INDIA PVT LTD

New Delhi

CRC Press
Taylor & Francis Group
Boca Raton London New York

CRC Press is an imprint of the
Taylor & Francis Group, an **informa** business

First published 2026
by CRC Press
4 Park Square, Milton Park, Abingdon, Oxon, OX14 4RN

and by CRC Press
2385 NW Executive Center Drive, Suite 320, Boca Raton FL 33431

CRC Press is an imprint of Informa UK Limited

British Library Cataloguing-in-Publication Data
A catalogue record for this book is available from the British Library

Print edition not for sale in South Asia (India, Sri Lanka, Nepal, Bangladesh, Pakistan or Bhutan).

ISBN: 9781041217541 (hbk)
ISBN: 9781041231172 (pbk)
ISBN: 9781003735427 (ebk)

DOI: 10.1201/9781003735427

Typeset in Times New Roman
by Mind Box Solutions, New Delhi

WPI

Contents

Preface

The book comprehensively reviews the significant researches in silk fibre. During recent years considerable research has been done on silk. The various properties of silk have been well explored. The book discusses about the researches related to mulberry and non-mulberry silks. Influence of different chemical treatments and silk process variables have been studied. Application of techniques like SEM, XRD, FTIR, TGA, etc. has been effectively used in investigation of silk. The non-mulberry silks like muga, tussar and spider silks have been explored. This book has been written with the intention of stimulating future research in the area. Silk is not merely the queen of textiles, but also holds a good potential for application in many areas. I wish to duly acknowledge the authors whose noteworthy contributions have been included in the book for the benefit of readers and promote future research in the area. While care has been taken in proper presentation of contents, suggestions are invited from readers to further enhance the quality of the book.

I wish to express my gratitude to our beloved chairman Dr. P.N. Ravi, CEO Dr. Anusha, and Principal Dr. G. Mohan Kumar for their moral support and encouragement.

Dr. N. Gokarneshan

1

Copolymerization method of weighing silk

Summary: The grafting of silk with methacrylic acid (MAA) and methyl methacrylate (MMA) helps to avoid the weighing of silk with conventional tin salts. An increase in the 'graft add-on' results in gradual reduction of dye absorption of grafted silk. When the weight is increased to 70%, the handle is not affected significantly but can be used for dyeing with acid, basic and reactive dyes with about 10–20% lower color strength, based on the class of dye and the sequence of grafting and dyeing. The dyeing followed by grafting with reactive dyes exhibits desirable properties and is suitable for weighting of silk on industrial scale. There is a reduction in the moisture content of the grafted silk, accompanied with marked increase in the stiffness and drapability. The high percentage of rigidity is confirmed by the increase in crease recovery angle on polymer add-on. Also, the damage on repeated laundering is more for degummed silk as against that of grafted fibre.

1.1 Introduction

The raw silk is composed of silk fibroin and silk sericin, which are of protein origin. It loses 20–25% of its actual weight due to removal of sericin during degumming process. As silk is sold on weight basis, it becomes important to make up this loss in weight. The weighing of silk helps to compensate the weight loss. There are some practical constraints associated with the mineral weighing technique, which is a traditional method of weighing silk. One is that weight increase is limited to 25–30% and the other is that the insoluble salts are partly trapped in the fibre and deposited on the fibre surface; these are washed out gradually in the subsequent washing during use. In addition, tensile properties and fibre durability during laundering as well as thread brittleness and proneness to slicing are some of the drawbacks. As compared to the traditional method, the weighing of silk with synthetic polymer by particular graft polymerization with certain monomers is becoming increasingly popular [1,2]. Scientists [3–7] have studied the grafting of vinyl polymers onto silk fibre and tried to define the structural analysis of the grafted fibre. It has been observed that this weighing of silk through graft copolymerization is easy to carry out at least in the yarn form but can also be performed in the fabric form [8]. The present work was aimed at increasing the weight of silk yarn and fabric through graft copolymerization of MMA and MAA by chemical initiation method as the large polymer add-on is easy to obtain because of the

DOI: 10.1201/9781003735427-1

hydrophilic nature of the monomers. The parameters related to grafting such as monomer concentration, initiator concentration and reaction time have been optimized. The grafted fabrics have been assessed for the variations in rigidity, crease recovery fullness and handle. In order to know whether grafting should be carried out before, after and during dyeing, the dyeing behavior with acid, basic and reactive dyes has been investigated.

1.2 Silk degumming

It is seen that the gum removal from raw silk is very less (Table 1.1) up to 60°C, irrespective of duration of treatment. But during a temperature of 70°C, the gum is removed from the fibre, even at 30 min treatment time. When the treatment temperature is 80°C and duration is 1 hour, the gum removal appears to reach a more or less constant value; and further increase in treatment temperature and duration shows no significant improvement in the gum removal from the fibre [12]. The moisture content of the raw silk is found to be about 9.3%, and after degumming at 80–100°C these value decreases to a more or less constant value of about 7–8%, where removal of gum from the fibre is around 22–23%. The reduction in the moisture content in degummed fibre could be attributed to the fact that gum and other natural impurities present in the fibre before degumming absorb more moisture on the fibre surface, as compared with the pure silk after degumming. Therefore the grafting treatment has done on fully degummed fibre to achieve better and uniform heterogeneous chemical reaction.

Table 1.1 Influence of temperature and duration of treatment time on degumming of silk [12]

Temperature (°C)	Treatment time (min)	Degumming (%)
50	30	0.87
	45	1.02
	60	2.50
60	30	1.80
	45	1.90
	60	2.95
70	30	3.80
	45	10.00
	60	19.00
80	30	20.00
	45	21.00

Contd...

Contd...

Temperature (°C)	Treatment time (min)	Degumming (%)
	60	22.00
90	30	21.00
	45	22.00
	60	22.00
100	30	23.00
	45	23.00
	60	24.00

1.3 Silk grafting

The silk fibre grafting has been done in aqueous medium in the presence of different concentrations of MAA with initiators: potassium peroxydisulphate and cerric ammonium sulphate. Reducing agents such as sucrose, ferrous ammonium sulphate, and sulphuric acid have been added as reducing agent to prevent air oxidation and homopolymer formation, and to keep the pH acidic, respectively. The graft yield increases progressively for both the monomers with the increase in monomer concentration; and the extent of grafting during final stages of monomer concentration is lower (Table 1.2). The maximum graft add-on depends on the type of monomer used. In case of MAA, the gel effect is not so pronounced up to a monomer concentration of 4 mol/litre; and in the case of MMA, the maximum graft add-on is obtained at a monomer concentration of 1 mole/litre [12]. Beyond the optimum concentration of monomers at which maximum graft yield is obtained, most of the monomers seem to be physically absorbed on silk fibre and impede diffusion of the initiator inside the fibre, thereby lowering the graft yield. The findings also indicate that the grafting of silk with MMA happens at a lower concentration in comparison to that of MAA.

Table 1.2 Influence of monomer concentration on graft yield [12]

MAA conc. (mol/litre)	Graft add-on (with FAS)	MMA conc. (mol/litre)	Graft add-on (%)	
			Without FAS	With FAS
2	40	0.05	3	1.0
3	50	0.10	4	2.5
4	112	0.50	43	30
5	125	1.00	98	61
6	–	1.50	42	31
7	–	2.00	18	14

At greater monomer concentrations, the grafted yarns become sticky and lumpy, due to which the fibre cannot be restored despite washing with agitation. Ferrous ammonium sulphate was therefore added in the bath to reduce the competing homopolymerization reaction, which has a negative effect on the handle and fastness. The presence of little amount of FAS in the graft bath also reduces the graft yield. This may be due to the fact that it may reduce the homopolymer formation on the fibre and at the same time retain the homopolymer in suspension. Moreover, the silk fibres are free from any serious stickiness and the homopolymer can be removed from the fibre easily by washing, thus rendering decrease in true grafting.

1.4 Standardization of grafting reaction

The quantities of initiator, ferrous ammonium sulphate and glucose, have been standardized by varying the concentrations of each factor in order to achieve optimum graft yield. Even though slight grafting occurs even at lower initiator concentration, it implies that the formation of silk radicals can initiate grafting (Figure 1.1).

The maximum graft yield is achieved for the two monomers investigated: at the initiator concentration of 10×10^{-3} mole/litre for MAA and 15×10^{-3} mole/litre for MMA without gel formation. The enhancement of graft by increasing initiator concentration up to a certain level indicates that the initiator decomposes to yield primary free radicals [12]. These free radicals may participate in direct abstraction of hydrogen atom from the silk fibre backbone to yield a silk macroradical capable of initiating grafting and may also affect the termination with the growing polymer chains. The graft yield increases with the increase in FAS concentration up to 1.5×10^{-3} mole/litre and then decreases with the further increase in FAS concentration. At higher FAS concentrations, the agglomeration of monomer takes place, thereby reducing the graft yield. Although no increase in graft yield is observed in the presence of FAS, the silk fibre remain definitely free from any stickiness and consequently washing out of the homopolymer is comparatively easier, thereby representing the more or less true grafting. The graft yield increases with the increase in sucrose concentration up to 1.5×10^{-3} mole/litre for both the monomers, and beyond this concentration there is a considerable decrease in graft yield. In a similar manner, the extent of graft add-on at various temperature and treatment time was also studied for both the monomers. At lower temperature and treatment time, the reaction rate is very low and so also the grafting. At higher temperature and treatment time, grafting increases due to the quick initiation and bonding between the fibre site and monomer.

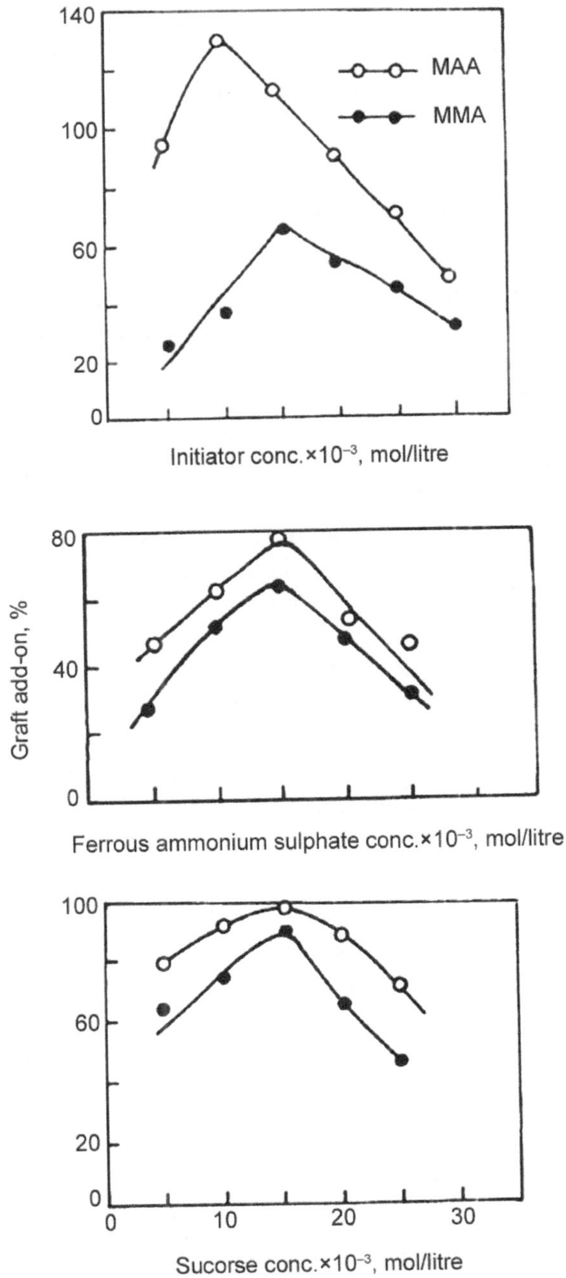

Figure 1.1 Standardization of grafting reaction [12]

However, above 70°C and 3 h treatment time, the increase in graft yield is not appreciable due to the chain transfer process since this graft formation acts as a diffusion barrier. Also, owing to the excessive gel and homopolymer formation, grafted fibre becomes very sticky, and as it is not removed completely, results in brittleness of the washed fibre.

1.5 Dyeability of grafted silk

The silk fibres have been degummed, grafted to different amount of add-on, and then dyed with acid, basic and reactive dyes. The amount of acid, basic and reactive dyes absorbed by silk grafted with MMA at various levels of add-on is depicted in Figure 1.2. Spectrometer has been used to measure the dye uptake by determining the color strength value after dyeing at 1% depth of shade for each dye chosen. It is observed that the amount of acid and basic dyes absorbed by the grafted silk fibre decreases with the increase in polymer add-on, and the extent of decrease in dye uptake depends on the class of dye used [12]; while the amount of reactive dyes is not much affected by the graft treatment. The decrease in the K/S value with the increase in graft add-on indicates that the dye absorption of grafted silk gradually decreases. In dyeing silk fibres with ionic dyes, the dye molecules are customarily attached to the amino and carboxyl groups by ionic bonds. The cause of decrease in the dyed absorption power is probably due to the blocking of polar or functional groups of the fibre chains by the monomer. A weight gain of 100%, which is considered to be actual weighting, causes no appreciable impairment of handle but shows dyeing with acid, basic and reactive dyes with about 10–25% lower K/S value.

In another set of experiments, the degummed silk fibre grafted with MAA were dyed separately with two acid dyes having two different sulphonic acid groups and two reactive dyes having different reactive systems in the dye molecules (denoted as graft → dye). The dyeing behavior has been determined. The degummed silk fibres (control) were also dyed (2% shade, o.w.f) under similar conditions. These dyed silk fibres were grafted with MAA under standardized conditions as mentioned above with a graft add-on of about 70%. The K/S values of these dyed samples are shown in Table 1.3. The results show that the degummed silk can be dyed with good depth of shade except for acid Scarlet 3R dye which contains three sulphonic acid groups in the dye molecule. This incidentally confirms that a weak acidic (acid) dye is more appropriate for silk than a strong acidic dye. The silk fabrics dyed with acid dye generally have low color fastness to washing, while the covalent bond formed by the reactive dye on silk substrate results in exceptionally good wash fastness properties. Results also show that the degummed silk fibres are more intensely dyed than grafted silk fibres. The relative intensity of dye (R.I.F) on degummed fibres (control) is denoted as 100. The R.I.F values of all the dyes decrease considerably when the grafted silk fibres are used for dyeing, and the extent of decrease depends on the class of dye and the sequence of grafting and dyeing. During grafting, the polar and functional groups of silk are blocked by the monomer, and therefore the availability of reactive groups for bonding with the dye molecule decreases.

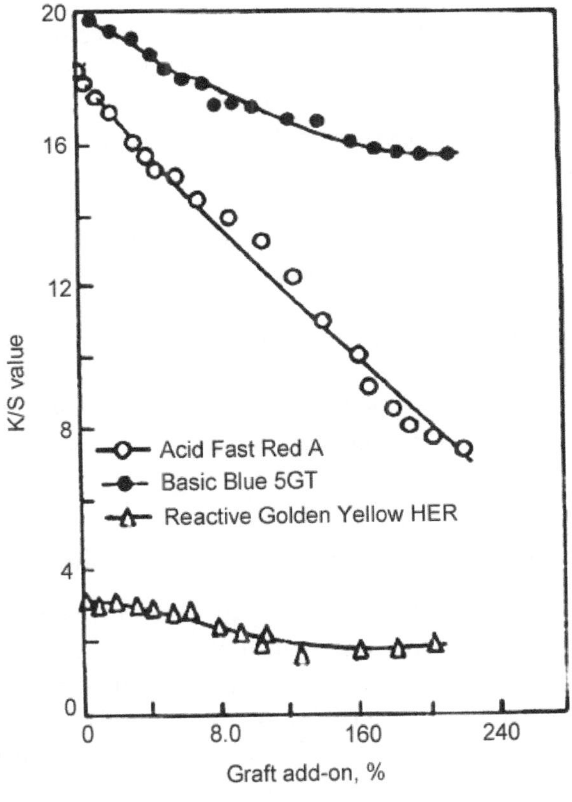

Figure 1.2 Dye uptake of grafted silk [12]

The other alternative way is grafting of dyed fibres by chemical initiation method (denoted as dye → graft). The dyeing behavior has been determined. The purpose of such inverted method is to study whether the dyes present in silk fibres have any adverse effect during grafting, i.e. the course of dyes must be capable of withstanding the grafting treatment condition involved. Results show that the acid dyes containing particularly three sulphonic acid groups in the dye structure have low K/S values, i.e. the dye structure is not stable in the grafting bath, resulting in almost negligible dyeing. In general the reactive dyes are not affected much during grafting reaction and their K/S and R.I.F. values are even greater than those obtained in graft → dye method. These dyes show uniform dyeing. It is, therefore, clear from this study that the silk fibres should be grafted first before acid dyes dyeing as the grafting after dyeing results in destroying the dye molecules, the extent of which depends on the number of sulphonic acid groups in the dye molecule. It is relevant to state that the simultaneous dyeing and grafting in one bath is not possible as practically no dyeing takes place due to instability of the dyes in the bath containing monomer and other chemicals.

The wash fastness properties of the grafted samples were also studied under mild conditions based on DIN 54005-A method. A good wash fastness grade was observed by the procedure of first dyeing and then weighing in comparison of first weighing and then dyeing. There is reduction in fastness with the increase in graft add-on. This can be explained either by mechanical entrapment of the dye after graft polymerization or by the fact that the relatively long weighing process acts like an after-washing step to improve the wash fastness. Thus it is clear that the sequence of treatment has a great effect on the wet fastness properties of the dyed fibres, but at the same time non-sensitive dyes are to be selected. It has been found that reactive dyes are more effective regarding this aspect.

1.6 Physical properties of grafted silk

The different physical properties of silk yarn grafted with MMA at various graft add-on percentages have been studied. When considering weighted silk, the reduction in color depth due to the increase in fibre diameter, or in other words the reduction in dye uptake, is influenced by the dye used and also percentage of grafting. Nearly half of the ungrafted fibre is penetrated by the acid dye and the tone of dyeing has been visually noted [12]. The changes with basic and reactive dyes were of the less order and the difference in tone was so small for a visual comparison of the dyeing that they lay on the borderline for clear discrimination. The results show that the rigidity is markedly increased by the grafting treatment. The reason behind the gradual loss in strength and hence the increase in retention ratio on increasing polymer add-on may be due to the removal of H-bonds between two polypeptide chains. The MMA polymer chains are cross-linked with each other when the polymer add-on is sufficiently high. Also the blocking of polar and functional groups of fibre chain due to grafting decreases the intermolecular attraction of polypeptide chain. The high percentage of rigidity was confirmed by the increase in the crease recovery angle on polymer add-on. The crease recovery angle increases slightly in warp direction and a little more in weft direction. Generally, the H-bonds are responsible for creasing of a fabric. Due to grafting, the H-bonds are broken and monomers are bonded with polypeptide chain. The presence of grafting material gives the required force to the fabric to return back to its original position, thus increasing the crease recovery. The moisture regain of the silk fibre is also influenced by the grafting reaction. A gradual reduction in moisture regain is seen with the increase in graft add-on, due to the hydrophobic properties of the grafted polymer molecules, which may be attributed to the loss in hydrophilic amino acid residues.

1.7 Influence of dyes on the physical properties of grafted silk

Investigations have been done on the influence of dyes on different physical properties of silk fabrics grafted before and after dyeing. There are quite significant differences in physical properties, which indicate the significant influence of grafting treatment on intrinsic textile properties of silk fibres. The breaking load and retention ratio of silk fibres are changed, depending on the sequence of dyeing and grafting. The breaking load of degummed silk fabric is about 20.7 kg and this value reduces to 10–14 kg for the grafted and dyed silk fibres [12]. If the samples are grafted first and then dyed, the breaking load is slightly lower than that of the samples dyed first and then grafted with a polymer add-on of about 38%. Blocking the polar groups of the fibre chain by grafting decreases the intermolecular attraction of polypeptide chain resulting in higher strength loss. The stiffness of the fabric was observed by measuring the bending length both in warp and weft directions under different sequences of application. The bending length in warp direction increases from 1.62 cm to 3.25 cm for degum → graft → dye fabric. The extent of increase in bending length is generally greater in weft direction than that in warp direction. The main reason for the increase in stiffness is that during grafting around the filament, the monomer can also cross-link with other monomer grafted on other filament. The filaments thus stick together and cohesiveness increases. In the case of weft, the number of filament is more than that in the case of warp and thus the bending length and stiffness are also higher for weft direction as compared to that for warp direction. The effect of grafting on drapability of the fabric was observed by the drape coefficient of degummed (control), the graft (69% add-on) → dye and dye → graft (38% add-on) fabrics. It is observed that the drape coefficient of the fabric increases from 0.279 (control) to 0.798 (69%) add-on. This remarkable increase in drape coefficient is that the drapability of grafted silk is improved sufficiently in comparison to that of ungrafted silk fibres. It may be because the fabric becomes stiff both in warp and weft directions, and hence increases the fabric folding property and drapability.

Upon repeated laundering tends to damage the surface of the silk fibres owing to abrasion. The bundles of fibrils, which are bound together by H-bonding mechanism in silk, are easily dissociated when mechanical forces are applied on wet silk, resulting in the lowering of breaking loads of laundered fibres. In the present work, the grafted fibres of different percentage of add-on were repeatedly laundered several times and the breaking loads of the laundered fibres were determined. The retention ratio (R) shows that the damage due to repeated laundering is comparatively more for degummed (control) silk fibres in comparison to that for grafted fibres. Similar effect was also observed by Kawahara et al. [11] in scanning electron micrograph of surface of silk grafted with MMA after laundering for 10 times. A network structure is formed on the grafted fibre surface due to the cross-linking between monomer polymer chains, which reduces the damage during laundering.

References

1. Brunet PC and Coles BC, Proc R Soc, London, Ser B, 187(1974)133.
2. Kawahara Y, Sen-I-Gakkaishi, 49(1993)621.
3. Samul S, Sahu G, & Nayak PL, Journal of Macromolecular Science; [A]Chem, 21(1084)725.
4. Tsukda M, Journal of Applied Polymer Science, 33(1988)2133.
5. Samal S, Sahu G, & Nayak PL, Journal of Applied Polymer Science, 29(1984)3283.
6. Nayak PL, Lenka S & Mishra S, Journal of Applied Polymer Science, 26(1981)3511.
7. Das RK, Basu D, Khan AK, and Banerjee A, Indian Journal of Fibres and Textile Research, 23(4)(1998)211.
8. Czemy AR, Uber AM, & Schindler W, Melliand Textilber, 71(12990)211.
9. Judd DB, & Wyszeci G, Colour in Business Science and Industry, 3rd Edition (John Wiley & Sons), New York, 1975.
10. Booth JE, Principles of Textile Testing (Butterworth Scientific, London), 1968.
11. Kawahara Y, Shioya M, & Takaku A, Journal of Society of Dyers color, 111(1985)382.
12 Karmakar SR, Mandal S, Das K, Sadhukhan D, Ghosh S, & Manna R, Indian Journal of Fibre and Textile Research, 27(2002)171.

2

Modification of silk fibre properties by graft copolymerization

Summary: Acrylamide has been grafted on non-mulberry silk through copolymerization using potassium permanganate-oxalic acid redox system in the presence of air. When the duration of reaction is 4 hours and treatment temperature is 50°C, the best results are obtained. The monomer (acrylamide), initiator (potassium permanganate), oxalic acid and sulphuric concentrations showing the best results are 2M, 15×10^{-3} M, respectively. The grafted silk fibres exhibit better thermal stability in comparison to ungrafted silk. Increase in grafting results in increase of the tensile strength and elongation of the grafted fibres. Also, increase in grafting reduces the water retention value. Hence, the water staining character of the silk is also minimized.

2.1 Introduction

Graft copolymerization is a proven technique for modifying silk fibres and fabrics through creation of branches of synthetic polymers which impart certain desirable properties to the backbone of desired materials without affecting their inherent characteristics. The chemical process involving vinyl monomers and different types of initiator systems result in modifications of the properties of natural protein fibres/fabrics, and are considered to be crucial. Such methods are very advantageous since they overcome some of the demerits related to them. Some vital changes in properties like thermal stability, photoyellowing, water repellency, and wrinkle recovery can be done by grafting [1,2]. Grafting of vinyl monomers onto natural macromolecules like silk, wool, and cellulose has received considerable attention during the last decade [3]. The graft copolymerization of silk fibres with vinyl monomers using different initiators is considered to be a powerful method for producing substantial modification in the physical, mechanical and morphological properties of the fibres [4]. These properties can be improved by graft copolymerization and/or by chemical modification techniques [5]. Most of the studies mentioned above have been carried out on mulberry and tussah silk fibroins. Literature survey has revealed that not much work has been reported on the graft copolymerization of muga silk (Antheraea assama) fibre, a precious natural fibre available in the northeastern part of the country. The silk fibre obtained from Antheraea assama, a multivoltine, sericogenic insect,

DOI: 10.1201/9781003735427-2

is popularly known as muga silk and has natural golden yellow hue. It is one of the most important and commercially valuable silk varieties available in northeastern India [6]. This silk also lacks some important performance properties. In the present work, graft copolymerization has been carried out onto muga silk fibre using vinyl monomer (acrylamide) and $KMnO_4$ oxalic acid redox system as the initiator in order to improve the quality of natural silk. The use of other monomer like MMA for grafting onto this non-mulberry silk fibre has already been reported [7]. The characterization of the grafted samples, obtained under optimum conditions, has been done by using Infra red and ultra violet spectra. Thermogravimetric, differential thermogravimetric and differential scanning calorimetry methods have been used to investigate their thermal behavior. The activation energy and frequency factor which are kinetic parameters, have also been evaluated [8].

2.2 Influence of temperature and time

The influence of reaction temperature on grafting of acrylamide onto silk fibres/fabrics has been investigated at four different temperatures ranging between 30°C and 60°C. Till a temperature of 50°C, there is steady weight increase and above it the weight gain rate gradually slows down. The higher activation energy has resulted in the increase in grafting (%). As the temperature of the reaction increases, the swellability of the fibre/fabric greatly enhances and as such the diffusion of the monomer from the solution phase to the fibre/fabric phase takes place [9]. The effect of reaction time on graft yield (%) was also studied for 1–5 h duration. The increase in grafting duration up to 4 h increases the graft yield (%), after which the grafting rate slows down. Thus the duration and temperature of reaction have strong influence on the grafting of acrylamide.

2.3 Influence of the concentrations of monomer, initiator, oxalic acid and sulphuric acid

The graft copolymerization of acrylamide onto silk fibre has been effected by varying the monomer concentrations from 0.5 mol/L to 2.5 mol/L, while maintaining other reaction conditions constant. As the monomer concentration rises up to 2M, the rate of grafting increases gradually and reduces subsequently. This might be due to the fact that at a certain monomer concentration, the combination of monomer probably takes place with silk molecules. When the concentration of polyacrylamide (PAAM) increases, the rate of their combination also increases faster than the rate of their combination with silk molecules due to gel effect [4]. The graft yield, graft conversion, graft efficiency and rate of grafting (%) at various monomer concentrations (acrylamide) in different durations of reaction are depicted in Figure 2.1.

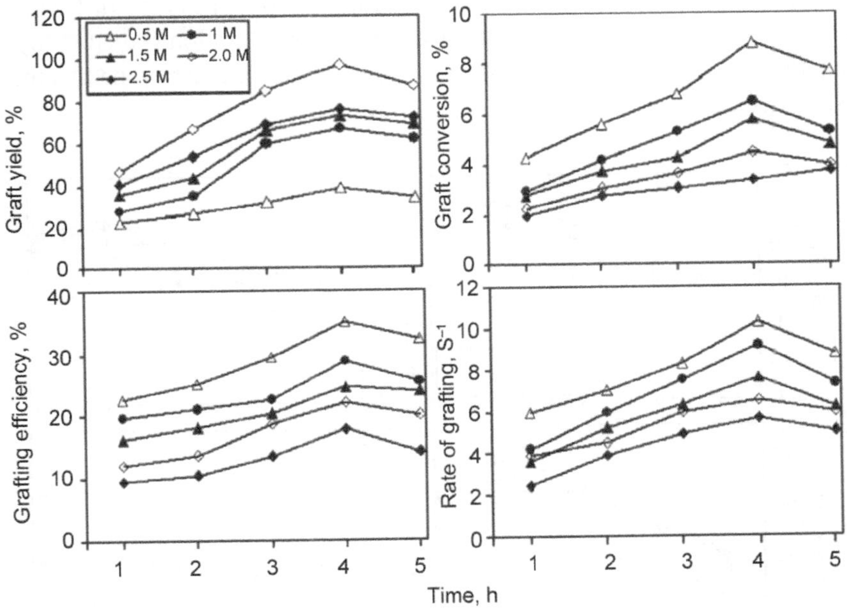

Figure 2.1 Influence of acrylamide concentration on graft yield, graft conversion, grafting efficiency and rate of grafting [19].

By means of changing the initiator ($KMnO_4$) concentration from 9×10^{-3} M to 17×10^{-3} M, the graft copolymerization has been effected. The rate of grafting (%) increases with the increase in initiator concentration up to 15×10^{-3} M and after that the rate of increase gradually slows down. As the concentration of permanganate increases, a large number of silk macroradicals are formed by the interaction of carboxyl-free radicals with the groups present in the silk backbone which initiate grafting, thereby increasing the graft yield. The graft yields decrease at higher concentration of oxidant, because the free radicals produced on the backbone of the silk fibre might oxidize to give rise to oxidation products [10]. The influence of $KMnO_4$ on the graft yield, total conversion, graft efficiency and rate of grafting in % is shown in Figure 2.2.

The concentration of oxalic acid ranges between 1×10^{-2} M and 3×10^{-2} M. At a critical concentration of oxalic acid, the maximum grafting (%) occurred, which corresponds to 2.5×10^{-2}. The grafting percentage reduces beyond this concentration. At higher concentration of oxalic acid, a considerable amount of manganese dioxide is used up in the side reaction and the rate of grafting decreases [11]. The influence of oxalic acid on graft yield (%), graft conversion (%), graft efficiency (%) is depicted in Figure 2.3. The concentration of sulphuric acid also varies from 13×10^{-3} M to 21×10^{-3} M in case of $KMnO_4^-$ oxalic acid initiator system. There is a critical concentration (19×10^{-3} M) of sulphuric acid at which maximum grafting (%) happens and beyond that there is a reduction in grafting.

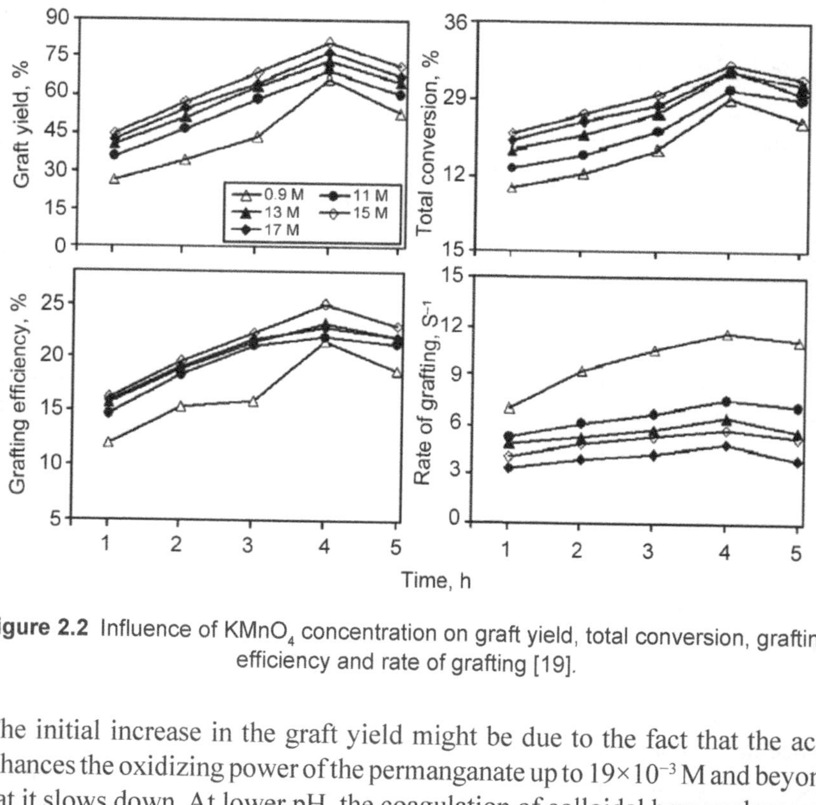

Figure 2.2 Influence of $KMnO_4$ concentration on graft yield, total conversion, grafting efficiency and rate of grafting [19].

The initial increase in the graft yield might be due to the fact that the acid enhances the oxidizing power of the permanganate up to 19×10^{-3} M and beyond that it slows down. At lower pH, the coagulation of colloidal homopolymer in solution and fibres might take place and as such the diffusion of both monomer and initiator to the fibre matrix reduces, resulting in decrease in graft yields. At higher concentration, manganese dioxide might react with acid, thereby producing oxygen which might inhibit the grafting process [12]. Figure 2.4 represents the effect of sulphuric acid concentration on graft yield (%), total conversion (%), graft efficiency (%) and rate of grafting in %.

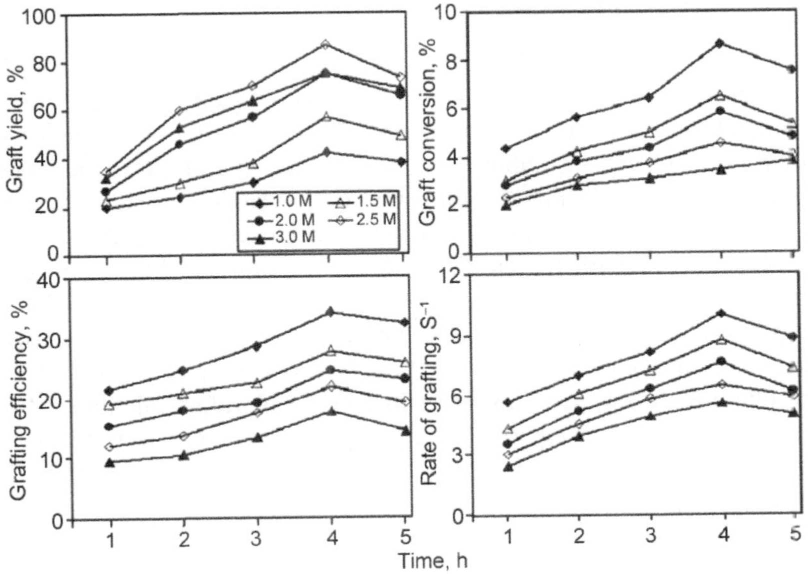

Figure 2.3 Influence of concentration of oxalic acid on graft yield, graft conversion, grafting efficiency and grafting rate [19].

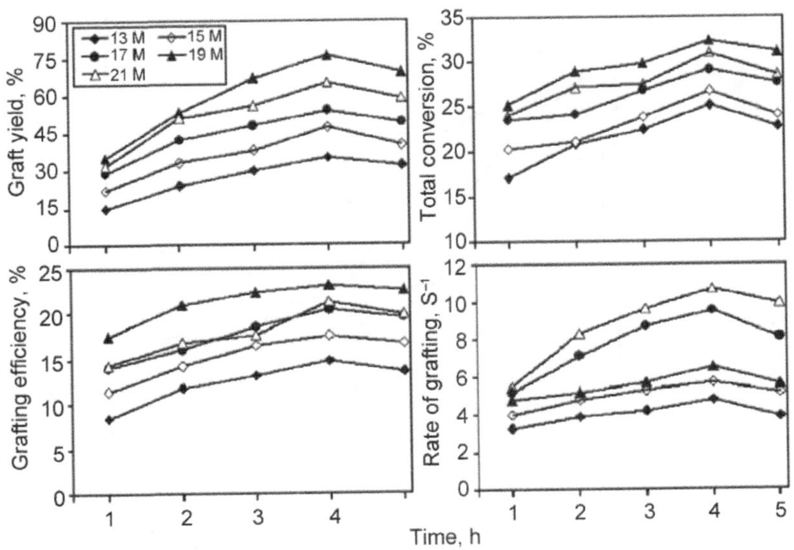

Figure 2.4 Influence of sulphuric acid concentration on graft yield, total conversion, grafting efficiency, and rate of grafting [19].

2.4 IR and UV studies

Both ungrafted and grafted silk fibres have been compared with regard to IR spectra. Characteristic bands at 1640 cm^{-1} for carbonyl group of amide I, 3420 cm^{-1} for NH– stretching vibration of amide II, 2860 cm^{-1} for $-CH_2$-(str.) and 3570 cm^{-1} for –CONH– group are exhibited by the grafted polyacrylamide silk fibre, thereby confirming the formation of acrylamide grafted silk fibre [13,14]. These bands were not present in the spectra of ungrafted silk fibre [19]. In the UV spectra for ungrafted silk fibre, one shoulder-form peak appears at 230 nm for –CONH- group of protein due to $n \rightarrow \sigma^*$ transition, a hump at 330 nm due to $n \rightarrow \pi^*$ transition of >C=O group of protein and another low peak at 430 nm due to $n \rightarrow \pi^*$ transition of group >CH-CO-NH- of silk protein (Figure 2.5). A hump is seen in the case of grafted silk fibre at 196 nm of $\sigma - \sigma^*$ transition owing to $-CH_2$-C< acrylamide group. A shoulder-form peak can be seen at 380 nm owing to $n \rightarrow \pi^*$ transition of protein group >CH-CO-NH- [15,16].

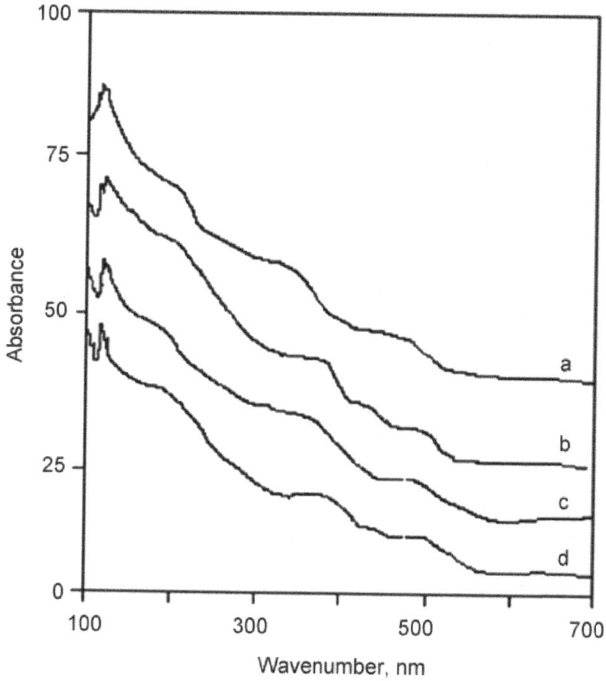

Figure 2.5 UV spectra of ungrafted and grafted silk fibres: (a) ungrafted, (b) 55% grafted, (c) 66% grafted and (d) 73% grafted [19].

2.5 Investigations of TG, DTG and DSC

The details relating to thermal (TGA and DTG) for grafted and ungrafted silk fibres have been obtained, at heating rates 20°C min^{-1} and 30°C min^{-1}. There are three stages in the thermal decomposition of fibres and are known as initiation, propagation and carbonization [17]. An initial mass-loss step at can be seen in all TG curves around 150°C, caused due to the removal of absorbed water. A major loss of weight is observed during the second stage in the case of ungrafted and grafted products (A-A3) at heating rate 20°C min^{-1}. The decomposition of protein starts at 150°C for ungrafted sample, which increases for sample A1 (180°C), A2 (185°C), and A3 (190°C), depending on the % increase in grafting, while in the third stage the decomposition of rest of the polymers starts at 390°C for A and then it increases for A1 (430°C), A2 (440°C) and A3 (450°C). The weight loss (%) of the grafted fibre is found to be less than that of the ungrafted fibre, which shows increase in the initial, maximum and final temperatures of active decomposition with the increase in grafting (%) of the silk fibre at both the heating rates [19]. The thermal behavior of grafted and ungrafted was studied with the help of DSC at heating rate 20°C min^{-1}; the thermograms are presented in Figure 2.6. The thermal analysis data are given in Table 2.3. In case of ungrafted one, the first broad endotherm observed below 100°C is due to the evaporation of water. At temperature range of 234°C and 297°C, two minor and broad endothermic (shoulder form) transitions are seen, and subsequently at temperature of 370°C a prominent endothermic peak is observed.

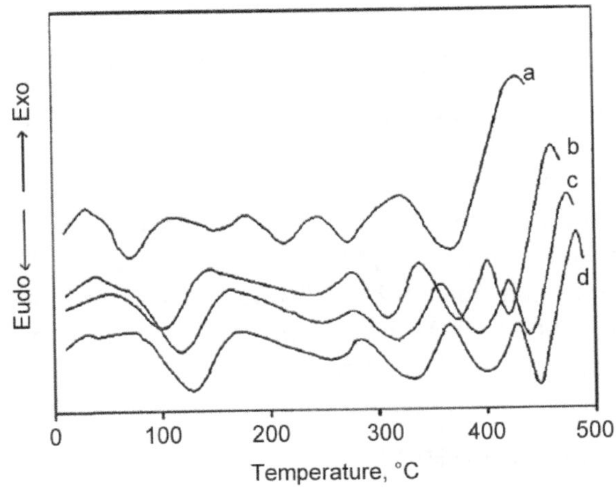

Figure 2.6 DSC curves of (a) ungrafted, (b) 55% grafted, (c) 66% grafted and (d) 73% grafted silk fibroin [19].

With the increase in grafting (%), the endothermic peaks shift towards higher temperature, due to evaporation of water as evident from the DSC data shown in Table 2.3. The minor peaks (shoulder form) shift towards 326°C, 333°C and 342°C for the fibres due to enthalpic change, and another endothermic peaks are observed in each of the grafted fibres at 388°C (55% grafted), 392°C (66% grafted) and 409°C (73% grafted). The findings from the DSC studies suggest that the endothermic peaks at 441°C, 445°C and 459°C for the fibres could be attributed to the presence of acrylamide polymer in the silk fibre. The glass transition temperatures (T_g) for the grafted silk fibroin have been studied.

2.6 Investigation of the kinetics

From the weight loss and temperature data, the kinetic parameters can be retrieved taking into consideration the factors: fraction decomposed, temperature, order of reaction, the frequency factor, the heating rate, gas constant, and activation energy. Evaluation has been done on order of reaction, activation energy, frequency factor, and standard error deviation using Fortran 77 computer program for the linear least square analysis with Gauss-Jordan sub-routine [19]. The data are found to fit well for a first order reaction. The values of activation energy and frequency factor for grafted products are found to be higher than that of the ungrafted one due to the high value of binding energy of grafted acrylamide monomer over that of silk. These values increase with the increase in molecular weights (increase in the percentage of grafting).

2.7 Evaluation of physical properties

2.7.1 Tensile properties

The tensile strength, elongation at break and stretch, with regard to ungrafted silk fibre, are seen to be 3.9 N, 16%, 30 mm; and in the case of grafted silk fibre, it is seen to be 5.6 N, 25%; for 50% grafted fibre, it is 41 mm; for 55% grafted fibres it is found to be 5.9 N, 27%, 43 mm; and for 73% grafted fibre it is 6.9 N, 34%, 47 mm [19]. As the percentage of grafting increases, the tensile strength, elongation at break and stretch also increase.

2.7.2 Water sorbency

The water sorbency of ungrafted and grafted silk fibres having various percentages of grafting has been found by measuring the water retention value adopting the technique above. In cases of fibres grafted to 50%, 55% and 73%, the water retention values are found to be 1.78 g/g, 1.69 g/g, 1.59 g/g, respectively; and in the case of ungrafted silk fibres it is 3.80 g/g [19]. In the case of AAm grafted, the water retention value decreases with the increase in grafting (%), and hence increases the hydrophobic nature of the fibre. The reduction in the water retention value could be attributed to the decrease in cohesive force of highly swollen fibres.

2.7.3 Water staining

Water is made to fall on the fabrics. The wetted fabrics (grafted and ungrafted) are sundried and then visually observed. There is no stain retention in the grafted silk fabrics [19].

References

1. Jian Ming Jiang, Hui Jun Zhu, Guang Li, Jun Hong Jin & Shen Ling Yang, Journal of Applied Polymer Science, 109(2008)3133.
2. Mansour OY, Nageib ZA & Basta AH, Journal of Applied Polymer Science, 43(1991)1147.
3. Jianquan Wang, Wenhui Wu, & Zhilui Lin, Journal of Applied Polymer Science, 109(5)(2008)3018.
4. Hongehun Li, Jincai Li, Yuetao Zhang, & Ying Mu, Journal of Applied Polymer Science, 109(2008)3030.
5. Tsukada M, Freddi G, Matsumara M, Shiozaki H, & Kasai N, Journal of Applied Polymer Science, 44(1992)799.
6. Hazarika LK, Saikia CN, Kataky A, Bordoloy S, & Hazarika J, Bioresource Technology, 64(1998)67.
7. Das AM, and Saikia C, Bioresource Technology, 74(2000)2013.
8. Coats AW and Redfern Jr, Nature, 68(1964)201.
9. Singha AS, Shama Anjali, & Thakur VK, Bulletin of Material Science, 31(2008)7.
10. Zhao Tie Liu, Chang'an Sun, Zhong Wen Liu, & Jian Lu, Journal of Applied Polymer Science, 109(5)2008, 2888.
11. Samal RK, Suryanarayanan GV, Das PC, Panda G, Das DP, and Nayak MC, Journal of Applied Polymer Science, 26(1981)222.
12. Mahanty N, Pradhan B, Mahanta MC, & Das HK, Journal of Macromolecular Science and Chemistry, A19(8)(1983)1183.
13. Barnwell CN, Fundamentals of Molecular Spectroscopy. 3rd Edition (Tata McGraw Hill Publishing Co. ltd., New Delhi), 1972.

14. Silverstein RM, Bossless G, Clayton Morill, & Terence C, Spectrometric Identification of Organic Compounds, 5th edition (John Wiley and Sons, UK) 1991.
15. Hirayana K, Handbook of Ultraviolet and Visible Absorption Spectra of Organic Compounds (Plenum Press data division, New York), 1967.
16. Scheinmann F, An Introduction to Spectroscopic Methods for the Identification of Organic Compounds (Pergamon Press, New York), 1970.
17. Isil Acar, Gulin Selda Pozan, & Saadet Ozgumus, Journal of Applied Polymer Science, 109 (2008) 2747.
18. Ali Habibi, & Ebrahim Vasheghani-Farhani, Journal of Applied Polymer Science, 109(2008) 3302.
19. Das AM, Chowdhury PK, Saikia CN, & Rao PG, Indian Journal of Fibres and Textile Research, 35(2010)107.

3

Silk dyeing with natural mordant

Summary : In the dyeing of silk with natural dyes, monascorubrin pigment extracted from the solid-state fermentation of monascus can be considered as an option. Monascorubrin pigment mordanted with terrae rare has been effectively used in dyeing of silk fabrics. The dye uptake and color strength value could be effectively increased by use of praseodymium chloride. The best adsorption capacity was realized with the simultaneous mordanting technique. It was found that the adsorption capacity decreased with increase in pH over the range. The optimal dye uptake occurred at pH 3. The dyeing rate of monascorubrin pigment alone was lower than for simultaneous mordanting with praseodymium chloride. The time of half dyeing reduced and the equilibrium dye uptake increased drastically for the mordanted sample. The use of mordant had enhanced color strength and colorfastness as well. The formation of covalent bond among the dye molecule, fiber and mordant could contribute to the good fastness of the mordanted samples. Monascorubrin pigment showed outstanding build-up for the silk fabric.

3.1 Introduction

Traditional natural pigments in general have their source in animal, plant or mineral origin [1–3]. For example, red pigment is obtained from madder, blue pigment from Indigo, and black pigment was obtained mainly from logwood. It has been recently known that natural pigment could also be obtained from a number of biological sources such as bacteria, fungi; and mildew could also be exploited for natural pigment production [4,5]. The application of natural pigments in natural coloration technology attracted increasing worldwide attention because these pigments are biodegradable and less harmful to humans and the environment. It could meet the consumer needs for safe and healthy life [6,7]. Nevertheless, dyeing with natural pigments has some problems such as low reproducibility, low dye exhaustion and poor fastness. In order to improve the dye uptake and color fastness, most of the dyeing processes were conducted by adding metal-based mordants [8]. These mordants comprised heavy metal ions such as copper, iron, aluminium, etc., which could cause environmental problems [9,10]. For these reasons, much effort had been spent to exploit ecological mordants. Terrae rare showed high compatibility with the environment and ecological safety, so it could be used as an eco-mordant for dyeing. Monascorubrin is extracted from the solid-state

DOI: 10.1201/9781003735427-3

fermentation of monascus. It is a water-soluble pigment with high purity and high color value. The stability of monascorubrin in the pH range 3–11 is good. The stability of the pigment against heat is also excellent. Monascorubrin has anticarcinogenic, antioxidative and hypolipidemic activities, which is usually used as an excellent food colorant for humans with its associated therapeutic benefits [11,12]. This chapter highlights the mordant dyeing of monascorubrin on silk fabric. Terrae rare has been used as mordant to decrease environmental pollution. The influence of various mordanting techniques (pre-mordanting, simultaneous mordanting, and post-mordanting) on the dyeing properties have been studied. The color fastness and kinetic factors of monascorubrin have also been studied.

3.2 Influence of mordanting techniques on the dye uptake and K/S values

Table 3.1 shows the influence of mordanting methods on the dye uptake and color strength values of silk fabrics. It can be observed from the table that the simultaneous mordanted sample exhibits higher depth of shade in comparison with other mordanting techniques.

Table 3.1 Influence of mordanting techniques on the dye uptake and K/S values [13]

Method of mordanting strength value	Uptake of dye (%)	Color
Pre-mordanting	80.83	3.4191
Simultaneous mordanting	87.61	4.8504
Post-mordanting	60.85	2.9735

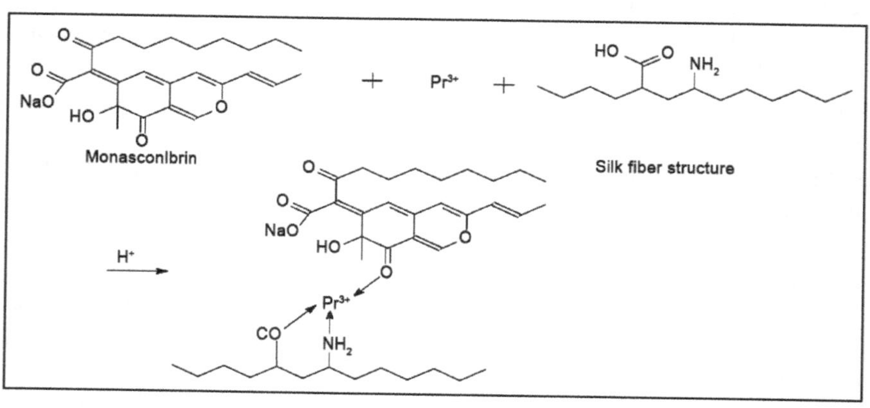

Figure 3.1 The mechanism of reaction [13]

Terrae rare mordant could mediate interactions between the pigment and silk fibre. Pigment Terrae rare ion complex would be formed on fibre molecules. The reaction mechanism is shown in Figure 3.1.

The interactions might be influenced by the mordanting method. Further research would be required to ascertain the reasons for the difference in color strength. Simultaneous mordanting has been used in all the successive investigations.

3.3 Influence of pH on the dye uptake and color strength (K/S)

Trials have been carried out to study the effect of pH on the adsorption capacity of monascorubrin on silk fabric, with different pH values using the simultaneous mordanting technique. Table 3.2 depicts the findings. The adsorption capacity reduces with increasing pH ranging 3–11. The possible mechanism for the effect of pH on adsorption of monascorubrin was likely to be ionic interactions of the pigment anions with the protonated amino groups on the silk fibre. When the pH of the solution was under the isoelectric point of silk (approximately 5.0), the silk would have more positively charged sites through increased amino group protonation than at higher pH values. A high negative potential has been observed. In other words the hydroxyl and carbonyl group form an electrostatic potential map of monascorubrin. Hence the adsorption process is enhanced owing to strong electrostatic attractions between the pigment and silk.

Table 3.2 Influence of pH on the dye uptake and color strength values [13]

pH	Dye uptake(%)	K/S
3	87	4.8
5	81	3.6
7	79	3.2
9	57	2.7
11	32	1.8

Rate of dyeing

The dyeing rate of monascorubrin pigment on silk fabric in the absence and presence of praseodymium chloride has been studied. Table 3.3 shows the kinetic factors of monascorubrin pigment are given.

The values of the time of half-dyeing were estimated from the experimental kinetic data. It was found that the dyeing rate of monascorubrin pigment alone was lower than for simultaneous mordanting with praseodymium chloride. The time of half-dyeing decreased and the equilibrium dye uptake increased drastically for the mordanted sample. The results confirmed that the mordanted samples had higher dyeing capacity than the unmordanted samples. The monascorubrin pigment dyeing rate has been mainly based on the affinity between monascorubrin pigment and silk fibres.

Table 3.3 Kinetic factors of dyeing [13]

Sample	Duration of half dyeing (min)	Exhaustion (%)
Unmordanting	31.78	56.90
Mordanting	29.49	87.61

In order to increase the affinity between the fibre and dyestuff, natural dyes frequently needed metallic mordant. The mordant increased the net positive charge of the fiber. The ionic interaction between the anionic pigment and silk fiber correspondingly increased in the presence of mordant. This electrostatic attraction plays an important role in enhancing the dyeability, which would increase the dyeing rate of monascorubrin pigment. The adsorption capacity with Monascorubrin pigment itself was smaller than that observed for the terrae rare complex. The anion of the monascorubrin pigment had a complex characteristic. Terrae rare ions can act as bridging ligands, and enable to form coordination linkages between the pigment molecule and fiber. Hence there has been an increase in the dyeing rate and equilibrium dye uptake.

3.5 Color fastness

Table 3.4 shows the results of washing fastness, light fastness, and rubbing fastness of the samples dyed with monascorubrin pigment in the presence and absence of mordant. It has been observed that the unmordanted samples show poor color fastness. The natural dyes are known for their problem of poor light fastness [13]. The bad washing fastness is due to the low affinity and fixation of monascorubrin pigment onto silk fibres. The good fastness of the mordanted samples might be attributed to the covalent bond formation between the dye molecule, fiber and mordant. Terrae rare ions have been to set pigments on fibers through formation of an insoluble compound which improves the wash fastness, rubbing fastness, and light fastness of the mordanted samples.

Table 3.4 Wash fastness, rubbing fastness and light fastness of the samples [13]

Sample	Washing fastness (grade)		Rubbing fastness (grade)		
	Staining	Fading	Wet	Dry	Light fastness
Unmordanting	3	3	3–4	3–4	2–3
Mordanting	4	4	4	4	3–4

Dye concentration – 2% o.w.f.; pH 3; bath ratio – 1:30; praseodymium chloride concentration – 2% o.w.f.

Build up characteristic

The build up curves of monascorubrin pigment on silk fabric have been studied. It is evident that the equilibrium dye concentration on the fibres increased with the initial dye concentration [13]. The equilibrium dye concentration in the fibres is a linear function of the dye concentration when the initial dye concentration is below 10% o.w.f. As the initial dye concentration went above 10% o.w.f, the slope of the build curve gets reduced. The finding shows that the monascorubrin pigment has excellent build up for the silk fabric.

References

1. Dotan Y, and Baslar S. Plants used as natural dye sources in Turkey. Econ Botany, 2003, 57, 442.
2. Chairat M, Bremner JB, and Chandrapromma K. Dyeing of cotton and silk yarn with the extracted dye from the fruit hulls of mangosteen. Garcinia mangostona Linn, Fibers Polymers, 2007, 8: 613.
3. Mongkholrattanasit R & Weiner J. Dyeing and fastness properties of natural dyes extracted from eucalyptus leaves using padding techniques. Fibers Polymers, 2010, 11:346.
4. Unagul P, and Wongsa P. Production of red pigments by the insect pathogenic fungus Cordyseps Unilateralis BCC 1869, Journal of industrial microbiology and biotechnology, 2005, 32:135.
5. Kobrakov KI, and Stankevich GS. The search for effective new dyes for chemical fibres. Fibre chemistry 2005, 37:84.
6. Lee DK & Cho DH. Fabrication of nontoxic natural dye from sappan wood, Korean journal of chemical engineering, 2008, 25:354.
7. Mati E, and de Boer H, Contemporary knowledge of dye plant species and natural dye use in Kurdish autonomous region, Econ Botany, 2010, 64:137.

8. Mansour Heba F, and Hefferman S. Environmental aspects on dyeing silk fabric with sticta coronate lichen using ultrasonic energy and mild mordants, Clean technology environment policy, 2011, 13:207.
9. Lee DK, Lee JH, and Shin HY, Fabrication of non toxic natural dye from sappan wood, Korean journal of chemical engineering, 2008, 25:354.
10. Lee YH, and Kim HD, Dyeing properties and color fastness properties of cotton and silk fabrics dyed with Cassia tora L, Fibers Polymers 2004, 5:303.
11. Mohammed MS, and Mohammed R, Enhancement of red pigment production by Monascus purpureous FTC 5391 through retrofitting of helical ribbon impeller in stirred tank fermenter, Food bioprocess technology, 2009, 10:214.
12. Babitha S, Julio C, and Pandey A, Effect of light on growth, pigment production and culture morphology of Monascus purpureous in solid state fermentation, World journal of microbiology and biotechnology, 2008, 24:2671.
13. Yanchun L, and Gang B, Textile research journal, 82(2), 203, 2011.

4

Effect of cooking and other treatments on the quality of raw silk

Summary : The cooking temperature and time and adjustment temperature profiles have considerable effect on the reeling characteristics, at standard temperature and time, high permeation temperature and time, and low permeation temperature and time. The reeling characteristics include groping end efficiency, reelability percentage, raw silk percentage, raw silk recovery percentage, waste percentage on silk weight, and degumming loss percentage on silk waste. On the other hand, parameters such as reeling tension, pelade weight and sericin dissolution percentage are not much affected by cooking treatments. The cocoon shell gets softened uniformly at a temperature of 95°C. But the boiling temperature oversoftens the sericin and entry suction phenomenon reduces the water permeation leading to lower performance in reeling. At a temperature range 93–60°C, the adjustment temperature is found to perform well in reeling characteristics. Owing to improper softening of sericin, the adjustment temperature of 93–80°C does not yield good results. This is due to the less difference between the adjustment temperatures. The reeling characteristics are also improved by the cooking time owing to proper softening of sericin. A cooking temperature of 95°C for 2 min time followed by adjustment temperatures of 93–60°C yields better reeling results in the case of multi-bivoltine cocoons. Also the complete cooking of hot air dried multi-bivoltine cocoons yields better silk recovery. The cocoon shell softens uniformly and the raw silk quality is improved considerably at cooking temperature of 95°C. The adjustment temperature does not show significant influence on quality characteristics because it enables maximum amount of water to permeate into the cocoons during complete cooking of Indian multi-bivoltine cocoons. The cooking time enhances the cohesion characteristics of raw silk due to the proper softening of sericin.

4.1 Introduction

A novel technique has been successfully developed for releasing the cocoon filaments from the cross over points between sericin layers adopting the process of complete cocoon cooking. The cooking process involves the following steps:

- Retting process wherein the cocoon shells are moistened with hot water.
- The air within the cocoon is expanded and expelled in order to create a pressure difference inside it by means of high permeation temperature.

DOI: 10.1201/9781003735427-4

- The cocoon is instantly treated with low temperature water by means of low permeation temperature in order to enable it to suck water through its layers.
- The water inside the cocoon is enabled to exit through its layers by means of cooking so that its inner layer of sericin is softened effectively.
- Adjustment in which the temperature of water is gradually reduced to enable the cocoon to absorb maximum amount of water.

The sericin in the cocoon shell is adequately softened during the above-mentioned stages in order to enable the cocoon filament to unwind smoothly without any break in the reeling process [1]. The cooking treatment becomes crucial as it decides the amount of absorbed water into the cocoon (semi-sunken/sunken system of reeling), waste production level, sericin swelling level adjustment and raw silk reeled quality [2–4]. Cooking technology was developed with a view to improve quality of raw silk and at the same time to reduce silk waste. A cooking technology involving pretreatment, cooking at 98°C for about 2 minutes with gentle boil followed by adjustment and gentle boiling was developed [5]. This method improved raw silk percentage and reduced the cleanness defects of raw silk. Systematic studies to understand the processes of cocoon cooking were carried out in Japan [6–14]. In these studies, the improvement in raw silk recovery was focused considering the concept of V-type conveyor cocoon cooking machine, in which the retting, permeation, steam cooking and adjustment temperatures could be controlled. This helps in setting the parameters for different quality cocoons. A cooking technique employing infrared heating in the cooking part was also discussed [15, 16]. The inner layer cooking is not achieved during permeation treatment, but it is achieved in cooking and adjustment treatments. The available literature clearly confirms that the complete cooking processes improve both the productivity and quality of raw silk. Nakagawa developed a cooking technology involving pretreatment, cooking at 98°C for about 2 min with gentle boil followed by adjustment and finishing. This method improved raw silk percentage and reduced the cleanness defects of raw silk. Shimazaki showed that the quality of raw silk is greatly influenced by cocoon cooking, reeling and allied operations of raw silk production [18]. He observed that the cleanness and neatness characteristics are improved by using suitable cooking temperature profiles [18]. Matsumoto studied the influence of drying temperatures on raw silk quality, i.e. neatness and cleanness characteristics [19]. Kinoshita et al. optimized the cooking methodology using multivariate analysis [20–22]. They inferred that the raw silk quality characteristics show significant improvement, depending on cooking temperature profiles. The entire cooking processes improve the productivity and quality of raw silk as confirmed by available literature. But, no study has been reported relating to the standardization of temperature profile for cooking and adjustment treatment suitable for hot air dried multi-bivoltine cocoons. Efforts have therefore been directed to investigate the influence of cooking temperature, time and adjustment temperature profiles with constant retting, and permeation treatments of hot air dried Indian multi-bivoltine cocoons on reeling and quality characteristics.

4.2 Technical details

Multi-bivoltine cocoons that are commercial available have been used. Water with quality characteristics comprising of pH value of 7, hardness of 80 ppm and alkalinity of 150 ppm have been used. The temperature gradient in cocoon drying has been maintained between 130°C and 90°C for 5-hour duration. The dried cocoons have then been retted and permeated. This has been followed by cooking and adjustment

4.3 Analysis of findings

The analysis of variance results of reeling and quality characteristics of Indian multi-bivoltine cocoons, based on cooking and adjustment temperature treatments, have been carried out. The mean results of the reeling and quality characteristics of the Indian multi-bivoltine cocoons treated with different cooking temperatures and durations have been determined. The mean reeling and quality characteristics results of the cocoons treated with different adjustment temperature profiles have been determined.

4.3.1 Analysis of variance results for reeling characteristics

The results of analysis of variance reveal that there is significant difference among cooking temperatures, cooking durations and adjustment temperature profiles with regard to reeling temperature characteristics, viz. groping end efficiency, reelability, raw silk, raw silk recovery, reeling tension, waste % on silk weight, pelade weight, sericin loss, and degumming loss (%) of silk waste [24]. From the results, it could be inferred that the cooking and adjustment temperatures are the significant factors which determine the reeling characteristics of Indian multi-bivoltine cocoons. The interactive influence among cooking temperature, adjustment temperature, and cooking duration do not have significant effects on all reeling characteristics, indicating that the effect follows the similar trend as performed individually.

4.3.2 Influence of cooking treatments on reeling characteristics

Grouping end efficiency

The ANOVA results show that the groping end efficiency is considerably influenced (5% level) by cooking and adjustment temperatures, while the time of cooking does not have significant effect. With the increase in cooking temperature from 93°C to 95°C, the groping end efficiency increases by 6.1%, while the groping end efficiency reduces by 5.8% at boiling temperature. The CD (critical difference) values at 5% level indicate significant difference between the cooking temperatures of 93°C, 95°C and 97°C with regard to groping end efficiency, which is attributed to the loose shell structure of the cocoons [24]. The results suggest that the cooking temperature of 95°C is highly suitable for Indian multi-bivoltine cocoons based on groping end efficiency. The cooking temperature of 97°C gives lower groping ends efficiency because of the less water penetration in the cocoons due to the entry suction phenomenon. It has been found that the adjustment temperature significantly influences the groping end efficiency. It is interesting to note that as the adjustment temperature increases, the groping end efficiency decreases by 6.3%. This is attributed to the fact that at higher difference in adjustment temperature profiles, more air pressure difference is created and more water permeates inside the cocoons. Thus, at 93–60°C adjustment temperature profile, more groping end efficiency is observed. With the increase in the cooking duration from 2 min to 3 min., the groping end efficiency reduces by 3%. The values at 5% indicate no significant difference in groping end efficiency.

Reelability

The reelability percentage is significantly influenced by cooking temperature at 5% level and cooking time at 1% level. There is no effect of adjustment temperature on reelability of multi-bivoltine cocoons. As the cooking temperature increases from 93°C to 95°C, the reelability is increased by 11.5%. A 6.7% drop in reelability is seen at boiling temperature, since at boiling temperature cocoons absorb less water due to the entry suction phenomenon. The cooking temperature of 95°C is observed to be quite suitable for multi-bivoltine cocoons [24]. The CD values at 5% level indicate that significant difference exists between the cooking temperatures of 93°C and 95°C with regard to reelability of the cocoons. This is because of the uniform sericin softening due to complete cooking of Indian multi-bivoltine cocoons. This is attributed to the pressure difference created by the adjustment temperature profiles. But, there is no significant difference among the adjustment temperatures with regard to reelability of Indian multi-bivoltine cocoons at 5% CD values. The increase in cooking time from 2 min to 3 min reduces the reelability of cocoons by 8%, which is significantly more as indicated by the CD values at 5% level.

Raw silk percentage

The yield of raw silk is significantly influenced by cooking temperature at 1% level and cooking time at 5% level. There is no significant difference due to shell structure of multi-bivoltine cocoons as seen from the adjustment temperature profiles. It shows that cooking temperature has a significant effect on the raw silk yield. It has been found that 95°C cooking temperature gives 13.64% raw silk. The raw silk percentage in Indian multi-bivoltine cocoons is higher till 95°C, whereas at boiling temperature the raw silk percentage decreases. This is because due to the thin shell thickness of Indian cocoons, they do not withstand the air pressure created at boiling temperature and result in less permeation of water during cooking. These cocoons break more during reeling and the repeated brushing reduces the raw silk yield. Hence, the cooking temperature of 95°C is suitable for better raw silk yield. The CD values at 5% level indicate that significant difference exist between the cooking temperatures with regard to raw silk percentage [24]. It is further observed that raw silk percentage increases as the cooking temperature increases up to 95°C, but it drops at boiling temperature in case of Indian multi-bivoltine cocoons. The CD values at 5% indicate no significant difference between the adjustment temperatures of 93–60°C and 93–70°C, but the adjustment temperatures of 93–60°C and 93–80°C show significant difference in raw silk yield. There is improvement in raw silk percentage with the increase in cooking time from 2 min to 3 min, which is significantly more as indicated by the CD values at 5% level. Hence, 2-min cooking time yields more raw silk percentage due to suitable softening of sericin, rendering effective unwinding of filaments from the cocoons.

Raw silk recovery

It has been found that the cooking temperatures significantly affect the raw silk recovery at 1% level. When the cooking temperature if raised from 93°C to 95°C, the raw silk recovery increase by 8%, while the increase in cooking temperature from 90°C to boiling temperature exhibits 1.5% reduction in raw silk recovery. The CD values at 5% level indicate that significant difference exists for raw silk recovery percentage between the cooking temperatures of 93°C and 95°C, but between cooking temperatures of 95°C and 97°C, so significant difference is observed [24]. This is because when the difference between the adjustment temperatures is large, more quantity penetrates inside the cocoons. In cooking process, as the energy available to remove this water is constant, complete removal of water cannot be achieved and hence the water penetration at 93–80°C reduces. At 93–60°C, this water is completely removed in cooking, resulting in maximum penetration of water in adjustment part. The CD values at 5% show no significant difference among the adjustment profiles of 93–60°C and 93–80°C with regard to raw silk recovery percentage. The raw silk recovery is enhanced by 4.2% with increase in cooking time. At 5% values of CD, a significant difference between the permeation durations is observed.

Reeling tension

The cooking temperature, cooking duration, and adjustment profiles do not affect the reeling tension [24]. Increasing the cooking time from 2 min to 3 min, increases the reeling tension by 6%. The CD values at 5% do not show any significant difference in reeling tension. It can be due to the uniform softening of sericin during cooking treatments for 2 min, resulting in better permeation of water in the cocoon shell from outside to inside.

Waste (%) on silk weight

The waste (%) on silk weight is significantly influenced by cooking temperature, cooking time and adjustment profiles at 1% level. The increase in cooking temperature from 93°C to 95°C decreases the waste (%) on silk weight by 9.6, while the increase in cooking temperature from 95°C to boiling temperature increases the waste (%) on silk weight by 5.9. It is due to over softening of sericin in outer layers at boiling temperature. The high permeation temperature of 95°C is observed to be quite suitable for multi-bivoltine cocoons with respect to waste (%) on silk weight [24]. The CD values at 5% level indicate that no significant difference exists between the cooking temperatures of 95°C and 97°C with regard to waste (%) on silk weight. This is attributed to the non-uniform treatment at higher adjustment temperature due to less water permeation, which, in turn, results in more dropping and repeated brushing, thereby increasing the waste percentage. However, the CD values at 5% show significant difference between 93–80°C and 93–70°C, but between 93–70°C and 93–60°C, no significant difference in waste (%) on silk waste is observed. The waste percentage increases by 0.6 as the cooking duration increases from 2 min to 3 min, which is not significantly more as shown at 5% level of CD values.

Pelade weight

The pelade weight is not influenced by cooking time and temperature and adjustment temperatures [24]. This may be attributed to better reelability of the multi-bivoltine cocoons and uniform softening of sericin during cooking. The CD values at 5% level also indicate no significant difference between the cooking temperatures with regard to pelade weight.

Sericin loss

The cooking and adjustment temperatures and cooking time do not considerably affect the sericin dissolution during cooking. This is attributed to sericin dissolution which is standardized in the permeation treatment, and there is no more notable dissolution during cooking treatments [24]. There is a decreasing trend in sericin dissolution by 4% as the cooking temperature increases from 93°C to 95°C. It is interesting to note that at 97°C, maximum dissolution occurs because of the oversoftening of sericin in Indian multi-bivoltine cocoons. The CD values at 5% level indicate that significant difference exists between 95°C and 97°C high permeation temperatures with regard to sericin dissolution percentage, whereas between 93°C to 95°C, no significant difference is observed. A cooking temperature of 95°C is found to be suitable for achieving less sericin loss from multi-bivoltine cocoons.

Degumming loss of silk waste

The degumming loss of silk waste is significantly influenced by cooking temperature at 1% level. Other parameters do not considerably affect degumming loss of silk waste. An increase in cooking temperature from 93°C to boiling decreases the degumming silk waste loss by 16.4%. The CD values at 5% level indicate that significant difference exists between the cooking temperatures of 95°C and 97°C with regard to degumming loss of silk waste percentage [24]. This is because at boiling temperature, more sericin dissolution takes place and hence the waste does not contain much sericin. The CD values at 5% do not show significant difference among the adjustment temperatures with regard to degumming loss of silk waste. The cooking time also reduces the degumming loss of silk waste by 3.6%. There is no significant difference between the durations of cooking at 5% values of CD.

4.3.3 Analysis of variance results for quality characteristics

The results of analysis variance indicate that significant difference exists among cooking temperature, cooking duration and adjustment temperature profiles related to quality characteristics, such as neatness, cleanness, tenacity, elongation, cohesion, and degumming loss of raw silk [24]. From the results, it could be inferred that the cooking and adjustment temperatures are also the most significant factors which improve the quality characteristics of

Indian multi-bivoltine cocoons. The interaction among cooking temperature, cooking duration and adjustment temperature profiles does not significantly affect the quality characteristics, indicating that it follows the similar trend as performed individually.

4.3.4 Influence of cooking treatment on quality characteristics

Neatness

The cooking temperature has a significant effect on the neatness of raw silk at 5% level. With the increase in cooking temperature from 93°C to 95°C, the neatness of raw silk is increased by 4.5%, while at 97°C the neatness is slightly lowered by 1.8%. It is owing to the fact that at higher cooking temperature the oversoftening of sericin occurs, resulting in the reduction of fibrils which reduces the raw silk neatness [24]. The CD values at 5% level indicate significant difference between the cooking temperatures with regard to neatness of raw silk, which is attributed to the softening of sericin during cooking process for multi-bivoltine cocoons. The results suggest that cooking temperature of 95°C is suitable for Indian multi-bivoltine cocoons based on neatness characteristics of raw silk. The lower and higher cooking temperatures reduce the neatness characteristics due to non-uniform softening of sericin in cooking process. Though the neatness is a racial characteristic to the extent of 80% as described by Shimazaki [23], the study shows that proper cooking could improve the neatness characteristic of multi-voltine cocoons to some extent. The adjustment temperature profile does not significantly influence the neatness of raw silk. However, the adjustment temperature profile of 93–60°C increases the neatness of raw silk by 1.7%, which is attributed to the fact that at higher adjustment temperature profile, the difference between cooking temperature and adjustment temperature profile reduces and this creates the lower air pressure difference so that the less water permeates inside the cocoons. When the permeation time increases from 2 min to 3 min, the raw silk neatness reduces by 1.4%. There is no significant difference seen in neatness of raw silk at CD values of 5%, which is due to water permeation into cocoons during the adjustment part.

Cleanness

The cleanness of raw silk is significantly influenced by cooking temperature at 5% level. The cooking time and adjustment temperature profile do not have any effect on cleanness of raw silk of multi-bivoltine cocoons. When the cooking temperature increases, the cleanness of raw silk also increases by 3.8%. It is interesting to note that the drop in cleanness of raw silk is observed at boiling temperature, which is because at boiling temperature due to entry suction phenomenon the water entry is restricted and this causes sericin softening. The cooking temperature of 95°C is observed to be quite suitable for multi-bivoltine cocoons [24]. The CD values at 5% level indicate that significant difference exists between the cooking temperatures of 93°C and 95°C with regard to cleanness of raw silk of the cocoons. It is observed that there is an increasing trend of cleanness of raw silk as the cooking temperature increases up to 95°C. However, the CD values at 5% did not show any significant difference among the adjustment profiles with regard to cleanness of raw silk. With the increase in the cooking time from 2 min to 3 min, the cleanness of raw silk reduces by 1.6%, which is significantly more as indicated by the CD values at 5% level.

Tenacity

The raw silk tenacity is not influenced by cooking, adjustment temperatures and cooking time, as its tenacity is quite high compared to other fibres. The tenacity is above the minimum level of 3.6 g/den at all the cooking temperatures used [24]. The CD values at 5% level indicate no significant difference between the cooking temperatures with regard to tenacity of raw silk. The CD values at 5% do not show significant difference in tenacity of raw silk. It is due to the uniform softening of sericin during cooking treatments for 2 min, resulting in better permeation of water into the cocoon shell.

Elongation

The raw silk elongation does not get affected by the cooking and adjustment temperatures and cooking time. For all cooking and adjustment temperature profiles, the elongation is above 19%. [24]. As the cooking time increases from 2 min to 3 min, the elongation increase by 0.2%. The CD values at 5% do not show significant difference in elongation with respect to cooking durations. This is due to the uniform softening of sericin during complete cooking treatments.

Cohesion

Raw silk cohesion is considerably affected by the cooking time at 5% level. The raw silk cohesion also increases by 13.8% with the increase in cooking temperature. The CD values at 5% level indicate that no significant difference exists between the cooking temperatures with regard to cohesion of raw silk [24]. However, the CD values at 5% do not show any significant difference between adjustment temperature profiles with regard to cohesion of raw silk. The cooking permeation time reduces the cohesion of raw silk by 14.9% as the duration is increased from 2 min to 3 min, which is significantly higher as indicated by the CD values at 5% level. This is due to the uniform softening of sericin in cocoon shell during 2 min permeation treatment.

Degumming loss of silk

The degumming loss of silk is considerably affected by the cooking temperature at 1% level. When the cooking temperature increases from 93°C to 95°C, the degumming loss of silk increases by 5.9%, whereas further increase in cooking temperature from 95°C to boiling decreases the degumming loss of raw silk by 7.3%, which is significant at 5% level as indicated by the CD values [24]. There is 4.1% increase in degumming loss of silk as the adjustment temperature profile increases from 93–60°C to 93–80°C, which is due to the sericin loss during complete cooking. The CD values at 5% do not show significant difference among the adjustment profiles with regard to degumming loss of silk. The cooking time also increases the degumming loss of silk slightly by 2.4%, which does not show significant difference as indicated by the CD values at 5%.

References

1. Kinoshita H. Bulletin of National Institute of Sericultural and Entomological Sciences, 19, 1997.
2. Kinoshita H, Watese H, and Sugenuma Y. Journal of Sericultural Science, Japan, 48(1979), 404.
3. Kinoshita H, Watese H, and Sugenuma Y. Journal of Sericultural Science, Japan, 49(1980), 352.
4. Kinoshita H, Kusama T, and Koeke R. Filature silk conference report, 32(1982), 10.
5. Nakagawa S, Journal of Sericultural Science, Japan, 14(1932), 43.
6. Shimazaki A, and Yoshizawa I. Filature silk conference report, 2(1952), 63.
7. Shimazaki A, and Yoshizawa I. Filature silk conference report, 3(1953), 119–125.
8. Shimazaki A, and Yoshizawa I. Filature silk conference report, 3(1953), 126–139.
9. Shimazaki A and Osan Takenaga, Filature silk conference report, 12(1962), 8.
10. Shimazaki A, and Sekijima K. Furuyama, Filature silk conference report, 12(1962), 12.
11. Shimazaki A, and Kono Uchida Saito. Filature silk conference report, 16(1966), 54.
12. Shimazaki A. Management engineering manual (Agricultural statistics association, Tokyo) 1973, 291.
13. Shimazaki A. Filature summer seminar report, 30(1977), 34.
14. Shimazaki A. Filature summer seminar report, 36(1983), 43.

15. Okamura G, Takegawa H, and Koiwae M. Filature silk conference report, 4(1954), 117.
16. Okamura G, Takegawa H, and Koiwae M. Filature silk conference report, 5(1955), 207.
17. Nakagawa S. Journal of Sericultural Science, Japan, 14(1932), 43.
18. Shimazaki A. Filature summer seminar report, 39(1986), 1.
19. Matsumoto T. Cocoon drying theory and practiced, PhD thesis, Tokyo University, Tokyo, Japan, 1984, 1.
20. Kinoshita H, Watese H, and Sugenama Y. Journal of Sericultural science, Japan, 48(1979), 404.
21. Kinoshita H, Tajima F, Miyajima T, and Simada M. Journal of Silk Science and Technology, Japan, 2(1993), 1.
22. Kinoshita H, and Tajima F. Journal of Silk Science and Technology, Japan, 3(1994), 7
23. Shimazaki A. Filature technology lecture (Sericultural Association, Tokyo) 1964, 89.
24. Hariraj G, and Somasekhar TH. Indian Journal of Fibres and Textile Research, 29(2004), 207.

5

Factors influencing quality and reeling performance of multi-bivoltine raw silk cocoons

Summary : The reelability, yield of raw silk from cocoons, and raw silk quality are directly influenced by the temperature and humidity during cocoon spinning. The reeling performance and raw silk quality are adversely affected by the high temperature and high humidity without air circulation. Cocoons spun under 27°C and 70% RH with air circulation of about 50 cm/s gave good reeling performance of cocoons and yielded raw silk of good quality. Temperature and humidity maintained during cocoon spinning will influence the cocoon (sericin) quality, particularly behavior of sercin during cooking, which, in turn, influences the reeling performance and quality characteristics of raw silk. This is the reason why cocoon quality, particularly reelability, from the same race deteriorates during rainy season (with high humidity) and hence affects the reeling performance and quality characteristics of raw silk. The reeling characteristics of cocoons and quality characteristics of raw silk are improved significantly with the adequate air circulation and good ventilation during cocoon spinning under high temperature and high humidity conditions.

5.1 Introduction

The agglutination force between the cocoon filaments at cross over points, the cooking condition of the cocoon and the weak spots along the length of the cocoon filament have a significant effect on cocoon droppings during reeling due to filament breakage. Also, the agglutination force between the cocoon filament and cocoon shell by proper softening and swelling of sericin determines trouble-free unwinding of the cocoon filament from the cocoon shell [1]. Uniform and proper softening and swelling of sericin depend upon the sericin characteristics, cocoon drying and cooking conditions [2,3]. The sericin characteristics appear to be influenced by climatic conditions during cocoon spinning. The studies have been conducted in Japan to observe the influence of temperature and humidity during cocoon spinning on reelability with bivoltine hybrid cocoons of their origin [4,5]. However, very little information is available relating to the effect of temperature and humidity maintained during cocoon spinning on other reeling characteristics, viz., average filament length, non-broken filament length, single cocoon filament denier, raw silk recovery, raw silk % and waste %, and yarn quality

DOI: 10.1201/9781003735427-5

characteristics, viz., neatness, cleanness, tensile properties and cohesion, of raw silk, and more so with Indian races. This chapter highlights the effect of temperature and humidity maintained during cocoon spinning on reeling characteristics and quality of raw silk of Indian multi-bivoltine cocoons.

5.2 Effect of temperature and humidity on reeling performance

Figures 5.1–5.4 depict reeling performance of multi-bivoltine cocoons spun under different temperature and humidity conditions. Cocoons spun at low temperature and low humidity (with air circulation) give significantly better reelability, average filament length, non-broken filament length, raw silk % and raw silk recovery in comparison with those of cocoons spun under high temperature and high humidity (particularly without air circulation)

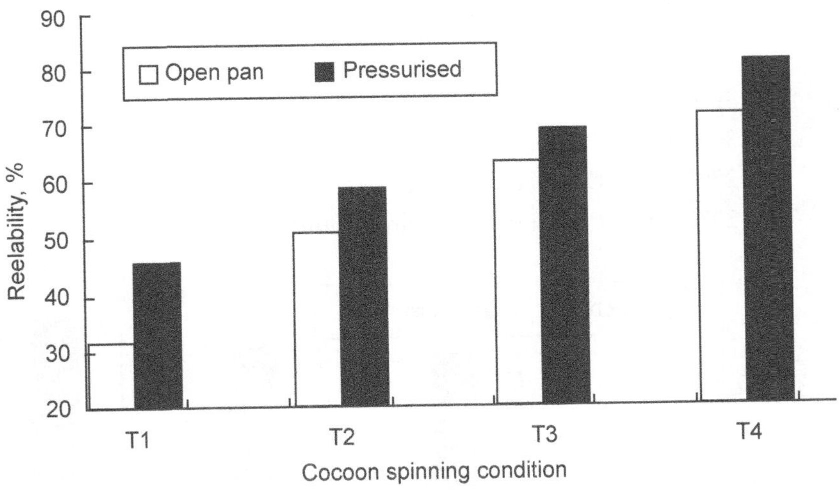

Figure 5.1 Effect of temperature, humidity, and air circulation on reelability of cocoons [11]

T1 – 30°C and 90% RH (without air circulation)

T2 – 30°C and 90% RH (with air circulation)

T3 – 27°C and 70% RH (without air circulation)

T4 – 27°C and 70% RH (with air circulation)

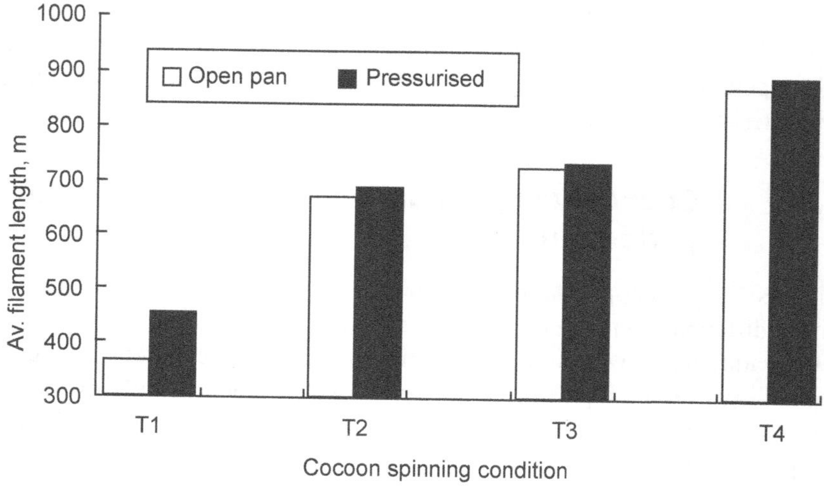

Figure 5.2 Effect of temperature, humidity, and air circulation on average filament length [11]

T1 – 30°C and 90% RH (without air circulation)
T2 – 30°C and 90% RH (with air circulation)
T3 – 27°C and 70% RH (without air circulation)
T4 – 27°C and 70% RH (with air circulation)

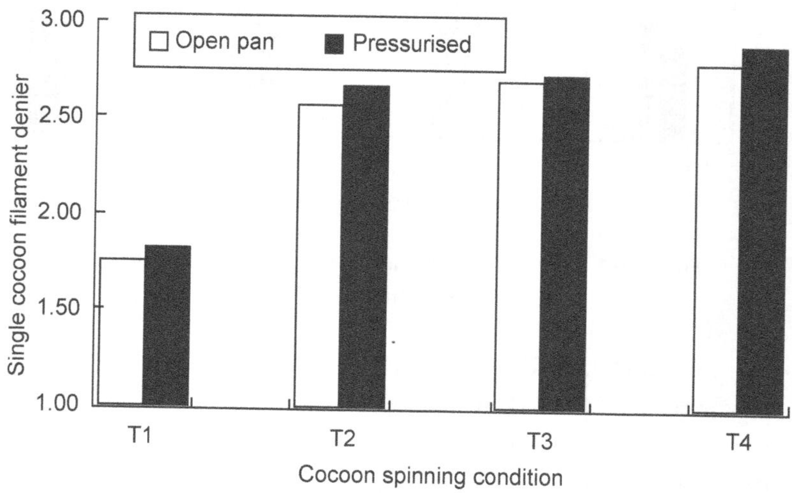

Figure 5.3 Effect of temperature, humidity, and air circulation on single cocoon filament denier [11]

T1 – 30°C and 90% RH (without air circulation)
T2 – 30°C and 90% RH (with air circulation)
T3 – 27°C and 70% RH (without air circulation)
T4 – 27°C and 70% RH (with air circulation)

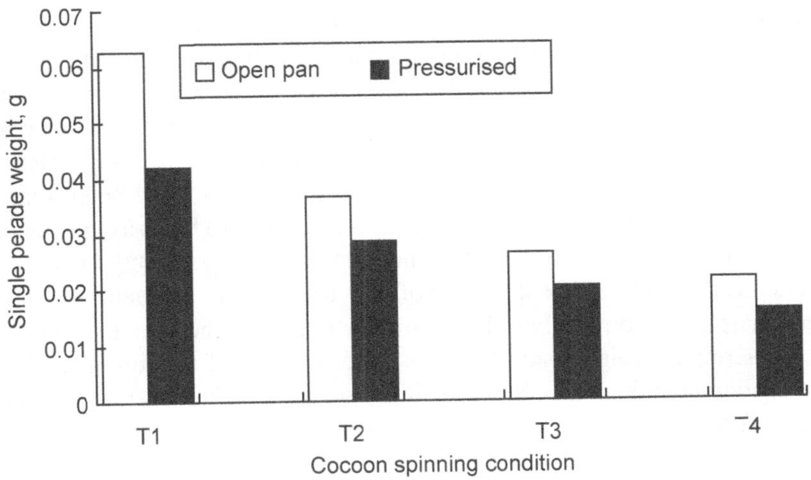

Figure 5.4 Effect of temperature, humidity, and air circulation on pelade weight [11]

T1 – 30°C and 90% RH (without air circulation)
T2 – 30°C and 90% RH (with air circulation)
T3 – 27°C and 70% RH (without air circulation)
T4 – 27°C and 70% RH (with air circulation)

Moreover, waste % is significantly less for cocoons spun with low temperature and humidity over that for cocoons spun under high temperature and humidity conditions. The effect of temperature and humidity on reliability, average filament length, non-broken filament length, single cocoon filament denier, raw silk recovery, raw silk % and waste % is very significant at 1% level. From the CD values, it is observed that the differences in the reeling characteristics, viz., reelability, raw silk % and waste %, between the treatments are significant at 1% level. It is also observed that the pressurized cooking method gives significantly better reeling performance and quality of raw silk as compared to open pan method. It is due to the better softening and swelling of sericin in all cocoon filament layers of cocoon shell in the case of pressurized cooking over that of pan cooking, where cocoons are cooked at only one temperature for lesser than necessary duration. In pressurized cooking, the cocoons are cooked following cooking principle that uses both steam and water as a cooking medium as compared to only water medium in open pan [2]. Results indicate that cocoon cooking has a significant role in achieving better reeling performance. However, it is observed that even with pressurized cooking the reeling performance results are not satisfactory in the case of cocoons spun under high temperature and high humidity conditions (particularly without air circulation) when compared with those for cocoons spun under low temperature and low humidity conditions. With pressurized cooking, the reelability decreases from 80.9% to 46%, the raw silk percentage decreases from 15.05% to 4.62% and waste % increases from 13.4% to 185.46%, when the cocoon spinning conditions are changed from 27°C and 70% RH (without air circulation). It is to be noted that the reelability of cocoons has significant influence on productivity and quality of raw silk, whereas raw silk percentage and waste percentage decide the economics of filature. The influence of high temperature and humidity during cocoon spinning on reeling characteristics is due to the structural changes in the sericin. A relationship between water content of cocoon layer during the cocoon spinning stage and reelability of cocoons has been found. Their study with bivoltine hybrid cocoons of Japanese origin reveals that the water content in the cocoon layers during cocoon spinning is significantly higher when the humidity is more as compared to that when the humidity is less. Correspondingly, the reelability of the cocoons is poor at high humidity. They have pointed out that the water content of the cocoon layer should be below 20% during cocoon spinning to obtain good reelability of cocoons. Study indicates that more water content (because of delayed evaporation) in the cocoon shell layers during cocoon spinning reduces the reelability of cocoons. Thus, it may be inferred that the low rate of evaporation of water content from cocoon layers during cocoon spinning would influence the structure of sericin. Therefore, to achieve good reeling performance, it is essential to remove the water present in the cocoon filament quickly. It is to be noted that after mounting, the silkworms drop their last faecal pellets and urinate. Urination is completed within one day after mounting. On an average, each silkworm drains off about 0.4 ml urine before spinning. It also increases the humidity in the cocoon-mounting chamber besides wetting the mountages/mat or papers. During cocoon spinning at high humidity, the water present in the spinning solution, silkworm urine and silkworm faeces gets slowly evaporated, thereby increasing the crystallinity of sericin. Solubility of sericin in water is found to be correlated with the structure of sericin and it decreases when the sericin molecules are transformed from random point to β structure and crystallized state [3]. This transformation is influenced by the moisture content of sericin during spinning [6]. From the X-ray diffraction pattern study on the crystallinity of cocoon sericin spun in the atmosphere of various relative humidity, Kataoka reported that as the relative humidity increases during cocoon spinning the crystallinity of sericin increases [6]. Ueda et al. reported that when the cocoon are spun under high temperature and humidity and low air flow, the sericin sticks firmly on the surface of filament and as a result, the solubility of sericin reduces [5–7]. Hence, the filament unwinding tension increases, resulting in excessive dropping of the cocoons during reeling which significantly reduces reelability. From the above discussion, it may be inferred that the structural changes that occur in sericin due to high humidity during cocoon spinning (particularly without air circulation) reduces the solubility of sericin and increases the agglutination force between the cocoon filament and cocoon shell. This, in turn, results in under/ uneven sericin swelling and softening during cooking and hence affects the reelability of cocoons severely. This is the reason why cocoon quality, particularly reelability, from the same race deteriorates during rainy season (with high humidity) and hence affects the reeling performance. Very poor reelability in the case of cocoons spun under high temperature and high humidity increases the waste percentage significantly because of frequent brushing and hence reduces the average filament length, raw silk recovery and raw silk percentage significantly (Figures 5.1–5.4). It is observed from Figure 5.3 that the linear density of single cocoon filament reduces from 2.90 denier to 1.83 denier, when the cocoon spinning conditions are changed from 27°C to 70% RH (with air circulation) to 30°C and 90% RH (without air circulation). Significant reduction in linear density of single cocoon filament at high temperature and high humidity during cocoon spinning is done due to the significant reduction in reelability of cocoons at this condition. Frequent brushing of cocoons because of poor reelability increases the brushing waste, leading to the removal of more number of outside filament layers which have coarse filament denier. Because of this, the average filament length and linear density of single filament reduce significantly in the case of cocoons spun under high temperature and high humidity. Figure 5.4 shows that the pelade waste is significantly on higher side in the case of cocoons spun under high temperature and high humidity as compared to that of cocoons spun under low temperature and low humidity. Higher pelade waste also contributes for reduction in raw silk % in the case of cocoons spun under high temperature and high humidity. Higher pelade waste in the case of cocoons spun under high temperature and high humidity may be attributed to the reduction in reelability of inner filament layers of cocoons. The moisture drying from the inner layer of sericin during cocoon spinning consumes greater time than those from the sericin of middle and outside layers, and more so at higher humidity. Hence sericin of inside filament layers will be very sticky owing to the structural changes and thus render it difficult to swell and soften. Consequently, the inside layers remain under/unevenly cooked during cooking, resulting in dropping of cocoons with higher pelade waste.

5.3 Effect of air current on reeling performance

The air current (circulation) leads to significant improvement in reeling characteristics of cocoons (Figures 5.1–5.4) [11]. This is due to the fact that forced air circulation/ventilation drives away the humidity from the cocoon-mounting zone and facilitates quick removal of water content from the cocoon layers during spinning, thereby helping in reducing the adhesive strength of sericin, depending upon the weather conditions. Because of this, there is a significant improvement in reelability, leading to significant improvement in raw silk recovery and raw silk %. The air circulation is crucial during cocoon spinning, specifically when the temperature and humidity are on the higher side for achieving better reeling characteristics.

5.4 Effect of temperature and humidity on raw silk quality

Figures 5.5–5.10 depict the effect of temperature and humidity on quality characteristics of raw silk is. In case of cocoons spun under low temperature and humidity conditions (particularly without air circulation), the neatness, low neatness, cleanness, tenacity and elongation characteristics of raw silk are significantly improved [11]. The influence of temperature and humidity on neatness, low neatness, cleanness, elongation and tenacity of raw silk is found to be highly significant at 1%. The results indicate that the pressurized cooking gives significantly better quality characteristics of raw silk as compared to open pan cooking. This could be attributed to the better softening and swelling of sericin of all layers of cocoon filament in the case of pressurized cooking as compared to that in the case of open pan cooking, where cocoons are cooked at only constant temperature for less time than required. In pressurized cooking, the cocoons are cooked using cooking principle [2].

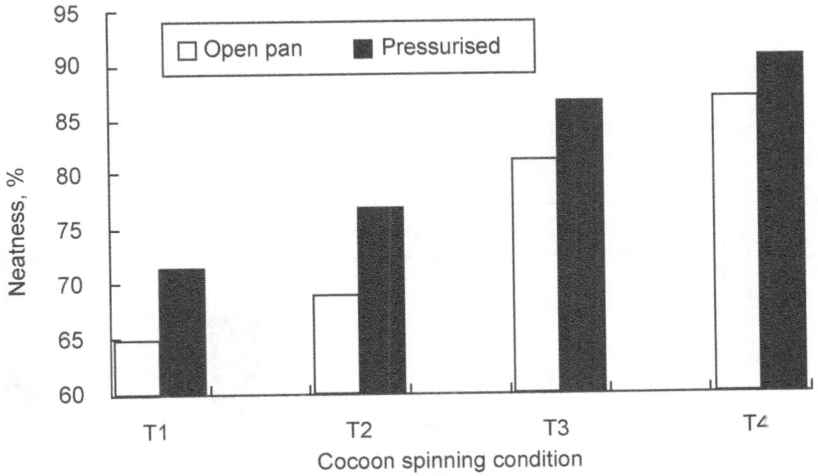

Figure 5.5 Effect of temperature, humidity and air circulation on neatness of raw silk [11]

T1 – 30°C and 90% RH (without air circulation)
T2 – 30°C and 90% RH (with air circulation)
T3 – 27°C and 70% RH (without air circulation)
T4 – 27°C and 70% RH (with air circulation)

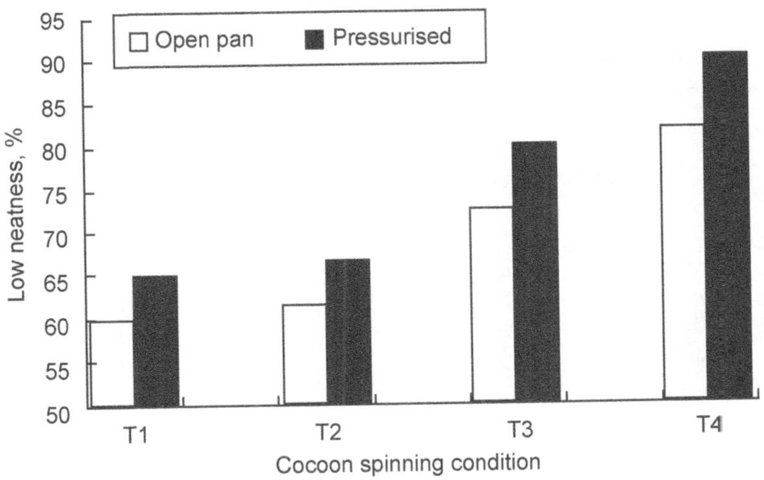

Figure 5.6 Effect of temperature, humidity and air circulation on low neatness of raw silk [11]

T1 – 30°C and 90% RH (without air circulation)
T2 – 30°C and 90% RH (with air circulation)
T3 – 27°C and 70% RH (without air circulation)
T4 – 27°C and 70% RH (with air circulation)

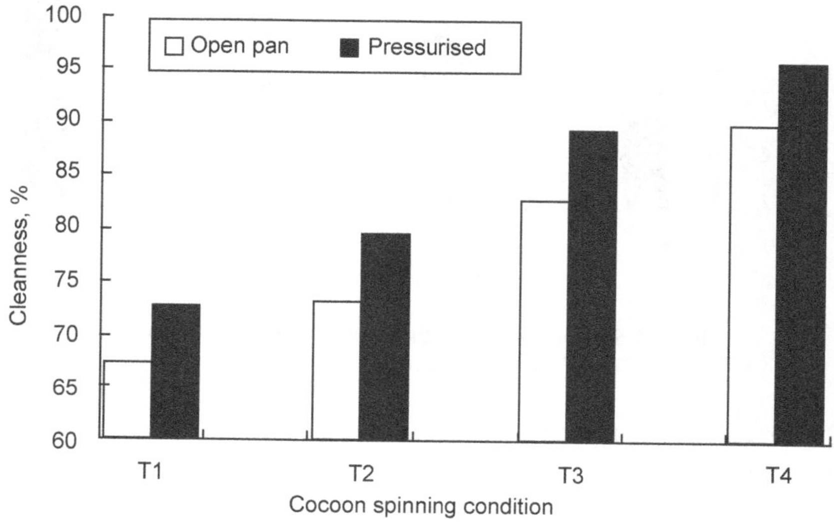

Figure 5.7 Effect of temperature, humidity and air circulation on cleanness of raw silk [11]

T1 – 30°C and 90% RH (without air circulation)
T2 – 30°C and 90% RH (with air circulation)
T3 – 27°C and 70% RH (without air circulation)
T4 – 27°C and 70% RH (with air circulation)

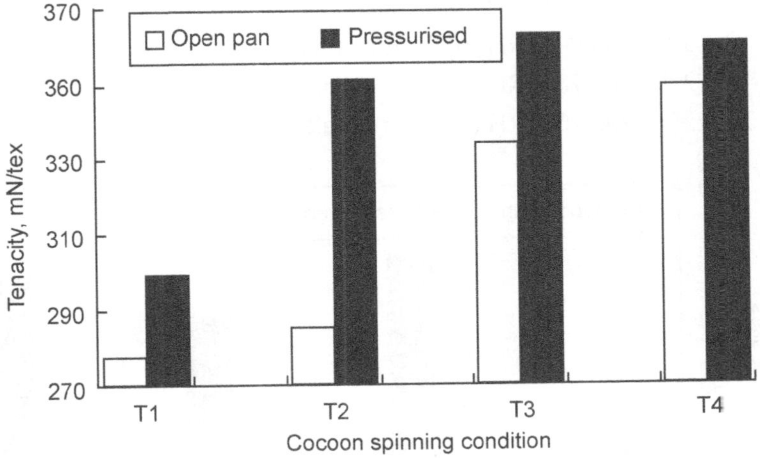

Figure 5.8 Effect of temperature, humidity and air circulation on tenacity of raw silk [11]

T1 – 30°C and 90% RH (without air circulation)
T2 – 30°C and 90% RH (with air circulation)
T3 – 27°C and 70% RH (without air circulation)
T4 – 27°C and 70% RH (with air circulation)

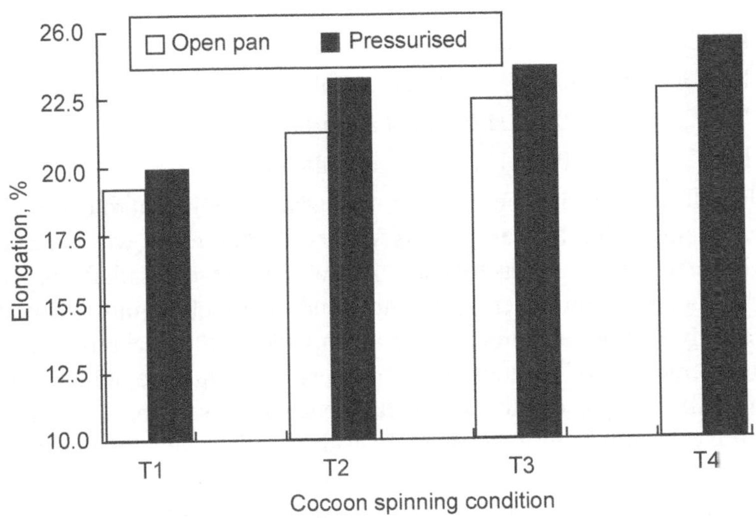

Figure 5.9 Effect of temperature, humidity, and air circulation on elongation of raw silk [11]

T1 – 30°C and 90% RH (without air circulation)
T2 – 30°C and 90% RH (with air circulation)
T3 – 27°C and 70% RH (without air circulation)
T4 – 27°C and 70% RH (with air circulation)

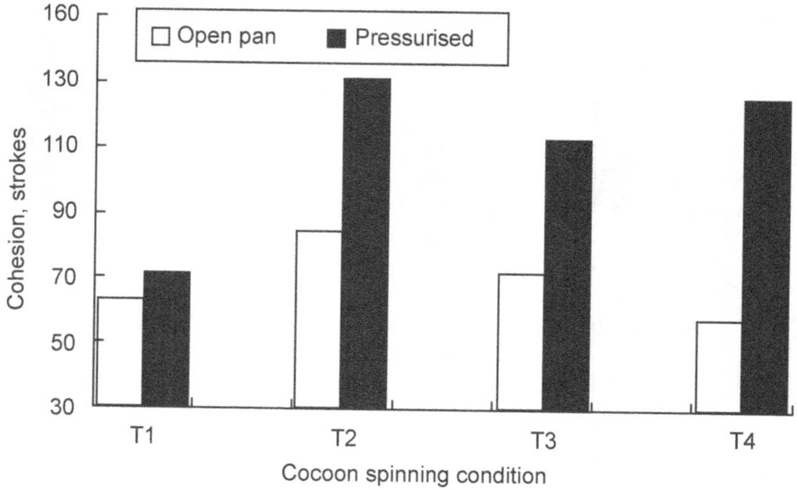

Figure 5.10 Effect of temperature, humidity and air circulation on cohesion of raw silk [11]

T1 – 30°C and 90% RH (without air circulation)
T2 – 30°C and 90% RH (with air circulation)
T3 – 27°C and 70% RH (without air circulation)
T4 – 27°C and 70% RH (with air circulation)

Results indicate that the cocoon cooking has a significant role in achieving better quality silk. However, it is observed that even with pressurized cooking, the quality characteristics of raw silk are not satisfactory in the case of cocoons spun under high temperature and high humidity conditions, particularly without air circulation as compared to the cocoons spun under low temperature and low humidity conditions. The influence of high humidity on raw silk quality is due to the structural changes in sericin. The structure of sericin has been related to stripping resistance (adhesive strength of sericin) of cocoon filament. From the measurement of stripping resistance of cocoon filaments and differential scanning calorimetric curves, study of cocoons spun under different temperature and humidity conditions, they pointed out that the stripping resistance of cocoon filament is weak in case of cocoons spun under low temperature and low humidity conditions because of the presence of only random coil structure in sericin. Whereas the stripping resistance and hence the adhesive strength of sericin is more in case of cocoons spun under high temperature and high humidity conditions, because sericin molecule in solution transforms from random coil to β structure due to the delayed drying of sericin. Increased adhesive strength of sericin in the case of cocoons spun under high temperature and high humidity reduces the solubility of sericin and increases the agglutination force between cocoon filament and cocoon shell. This, in turn, results in improper/uneven sericin swelling and softening during cooking thereby affecting the quality of raw silk. The relationship between the variation in the cooking conditions, agglutination force between the sericin cross over points and cocoon filament exfoliation tension which occurs when a cocoon filament is exfoliated from the cocoon layer during reeling, have been observed by researchers. During unwinding of a cocoon filament, there will be stretch in the filament, particularly when released from cross over points. This depends on the agglutination force between the cocoon filament and cocoon shell and the cocoon cooking conditions. If this force is very high, the filament may break, depending on the weak spots in the cocoon filament and cooking conditions, and hence the dropping of cocoon during reeling which reduces the reelability. Alternatively, the filament may be stretched beyond the yield point before it is released from the cross over points in the cocoon shell affecting the tensile properties of cocoon filament and hence the raw silk [9]. Because of very high agglutination force between cocoon filament and cocoon shell spun under high humidity conditions, the tenacity and elongation of raw silk are affected severely (Figures 5.8 and 5.9). This is the reason why cocoon quality, particularly, reelability from the same race deteriorates during rainy season (with high humidity) and hence affects the quality of raw silk. Because of β-structure (more adhesive strength) of sericin in the case of cocoons spun under high humidity conditions, it is difficult to achieve uniform and required softening and swelling of sericin of cocoon filament at cross over points in cocoon shell during cooking. It leads to uneven/underswelling and softening of sericin during cooking of cocoons, resulting in the occurrence of more neatness and cleanness defects such as loops, slugs, nibs and hairiness in the raw silk. This is the reason why the neatness, low neatness and cleanness characteristics of raw silk decrease very severely in the case of raw silk reeled from the cocoons spun under high humidity conditions (Figures 5.5–5.7). The multibivoltine cocoons due to loose structure and flossy nature are likely to produce more cleanness defects when they are unevenly cooked. The findings show that despite the neatness of the raw silk being influenced more by the racial character of the cocoons, the inherent cocoon filament neatness can be affected significantly during reeling the cocoons spun under conditions of high temperature and humidity. Figure 5.10 shows that the cohesion of raw silk reeled from the cocoons produced with high temperature and humidity using pressurized cooking method is more than that of silk reeled under other conditions. It is to be noted that the cohesion of raw silk is significantly influenced by racial sericin character and proper softening and swelling of sericin during cooking [10]. Results confirm that the adhesive property of sericin is more in case of cocoons produced with high humidity as compared to that of cocoons produced with low humidity. Interestingly the cohesion of raw silk reeled using open pan cooking with high humidity is significantly lower even though the sericin of these cocoons possesses better cohesive property. This is presumably due to the fact that hard and sticky sericin of cocoons spun under high humidity conditions cannot be softened easily and if such cocoons are improperly or undercooked with open pan, sericin will not be softened/swollen properly resulting in poor cohesion of filaments in the raw silk. For all the cases, the cohesion of raw silk reeled with pressurized cooking is good in all the conditions and significantly better than that of open pan cooking. It is due to the proper softening and swelling of sericin in the case of pressurized cooking.

5.5 Effect of air current on raw silk quality

The air current (circulation) significantly improves the quality characteristics of raw silk both in the case of high humidity and low humidity conditions during cocoon spinning (as seen from Figs. 5.5–5.10). It is because the forced air circulation/ventilation drives away the humidity from the cocoon-mounting zone and enables immediate removal of water content from the cocoon layers during spinning, thus aiding decrease in the adhesive strength of sericin, based on the weather conditions [11]. Because of this, sericin of these cocoons has less resistance towards cocoon cooking and hence facilitates uniform cooking, leading to the improvement in quality characteristics of raw silk. Results also indicate that the role of air circulation during cocoon spinning is very important, particularly when the atmospheric humidity is on the higher side, for achieving better quality characteristics. As per the interaction of CSTRI with reeling filatures and cocoon market during monsoon, particularly from July to October, average renditta increases by 0.5–1.0, productivity comes down by 20–25% and quality of raw silk deteriorates as compared to that in other seasons with the same race of cocoons because of the high humidity. This has significant influence on techno-economics of the filatures. Concrete efforts by means of integrated approaches are much needed to improve the cocoon quality and hence the reeling performance and raw silk quality in rainy season, through educating the farmers about importance of temperature and humidity management during cocoon spinning by using rotary mount ages, air circulation fans, heaters and by providing cross ventilation in rearing room, and so on. The use of dehumidifiers in the cocoon mounting room is very beneficial when the humidity is very high, as in continuous rains. The quality of cocoons gets improved by these efforts.

References

1. Takabayashi C, Nayak SV, and Tsuboi T, Journal of Sericultural Science Japan, 65(1996)278.
2. Omura T, Silk reeling techniques in tropics (Japan international cooperation agency, Tokyo, Japan) 1981, 61 & 125.
3. Komatsu KO, Bulletin of Sericulture Statistics Japan, 26(1975)135.
4. Akahane T, and Tsubouchi K, Journal of Sericultural Science Japan, 63(1994)229.
5. Kataoka K, Kobunshi Ronbunshu, 34(1977a)1.
6. Kataoka K, Journal of Sericultural Science Japan, 46(2)(1977)169.
7. Ueda S, and Suzuki K, Bulletin of Sericulture Statistics Japan, 103(1976)37.
8. Jan Zhu L, Arai M, and Hirabayashi K, Journal of Sericultural Science Japan, 103(1976)37.
9. Naik Subhas V, Takabayashi C, and Somashekar TH, Sericologia, 36(2)(1996)305.
10. Naik Subhas V, Takabayashi C, Tsuboi T and Somasekhar TH, Sericologia, 35(3)(1995)513.
11. Naik Subhas V, and Somasekhar TH, Indian Journal of Fibre and Textile Research, 29(2004)324.

6

Crystallite-shape ellipsoid of non-mulberry silk fibres

Summary : Investigations on mulberry and non-mulberry silk fibres using wide-angle X-ray diffraction studies of coarse and fine fibres reveal that even though there is not much change in the positions of X-ray reflections, there is a significant change in the shape of crystallite domains of non-mulberry silk fibres. It arises owing to a different composition and arrangement of alanine-glycine-serine residues in non-mulberry silk fibres. In both the cases of mulberry and non-mulberry silk fibres, the tenacity reduces with the crystal size increase. It necessarily highlights the fact that ordering in the lattice could be undesirable aspect in the textile industry. .

6.1 Introduction

The silk worm secretes a fibrous protein which becomes an external structure around its body, known as cocoon. The fiber is a semi-crystalline polymer that is widely used in the textile industry [1]. The commercial Indian silk varieties include mulberry, muga, tasar and eri. Hence it becomes necessary to understand the microstructural changes in these fibres, as these factors determine the strength and property of the fibres. Extensive research work has been carried out on the structure and property of mulberry silk fibres, but to a lesser extent on non-mulberry silk fibres [2–8]. Some of these studies report mechanical properties, particularly the load elongation behavior of raw mulberry silk fibres and to some extent of tassar and muga silk fibres. Rajkowha et al. have reported the tensile strain and recovery behavior of these silk fibres and their structural dependence. However, the variation in the crystallite-shape ellipsoid in these silk fibres has not been studied so far. Therefore, in this chapter, the wide angle X-ray scattering (WAXS) reported by Rajkowha et al. have been used to determine the exact behavior of crystal-size distribution in polymers like silk by employing three asymmetric functions and to find out a function which gives a better fit with the experimental profile and also the consistent microstructural parameters values in silk fibres [9]. These parameters are further used to compute the crystallite-shape ellipsoid. The microstructural factors are correlated with the reported physical parameters, such as tenacity and toughness.

DOI: 10.1201/9781003735427-6

6.2 Technical details

Fourier method of Warren has been used to determine the microstructural factors such as crystal size and lattice strain [10–12, 15]. The microstructural factors have been computed using equation that involved exponential, reinhold and lognormal distributions. A number of equations have been used to simulate the intensity profile by varying the necessary factors in order to obtain a good fit with the experimental profile.

6.3 X-ray investigations

Using various distribution functions, X-ray reflections of non-mulberry silk have been simulated with different distribution functions, [100] and [201]. The experimental and simulated X-ray intensity profiles obtained on the basis of different column length distribution functions for [201] reflection in mulberry fine and non-mulberry (tassar) coarse silk fibres are depicted in Figure 6.1(a–f) [15]. Such investigations have also been carried out for mulberry silk fibre to bring out the distinction between the two types of silk fibres.

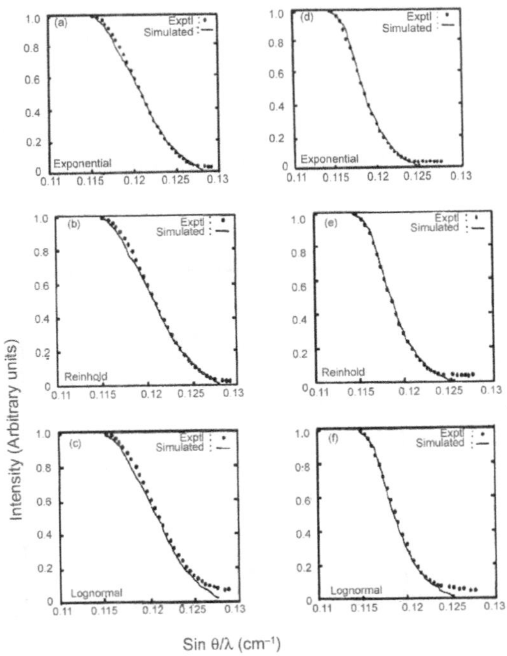

Figure 6.1 Experimental and simulated intensity profiles of [201] X-ray reflection [15] (a–c) Fine mulberry; (d-f) Coarse non-mulberry (tassar); (a, d) Silk fibres obtained with exponential; (b, e) Reinhold; (c, f) Lognormal column distribution functions.

It is observed that there is a relatively better fit between experimental and simulated profile in the case of exponential distribution function for the distribution of crystal sizes in polymers. Here, it is emphasized that the standard deviations in all the cases for the microstructural parameters are given as 'delta'. As exponential distribution function gives a better fit than others, the corresponding results obtained indicate following three important features:

(a) Surface weighted crystal size is more for non-mulberry silk fibres when compared to mulberry silk fibre.

(b) Crystal size decreases with degumming treatment, and

(c) Crystal size is more for coarse fibres.

The variation in lattice strain (g) lies between 1% and 4.5% in the case of exponential distribution for both coarse and fine fibers. From the obtained microcrystalline parameters, one can estimate the minimum enthalpy which defines the equilibrium state of micro-para crystals in silk fibre [15]. The relationship is $\alpha = (N)^{1/2}$, as given by Hoseman [16]. This value of α implies physically that the growth of paracrystals in a particular material is appreciably controlled by the level of lattice strain in the net plane structure. The estimated minimum value of enthalpy obtained on the basis of exponential distribution function lies between 0 and 0.12 for the coarse and fine non-mulberry silk fibres. Normally, for polymers this value lies between 0 and 0.2, which is the case in silk fibres. The variation in crystal size distribution along [201] direction obtained on the basis of various column length distribution function for eri fine fibres is shown in Figure 6.2.

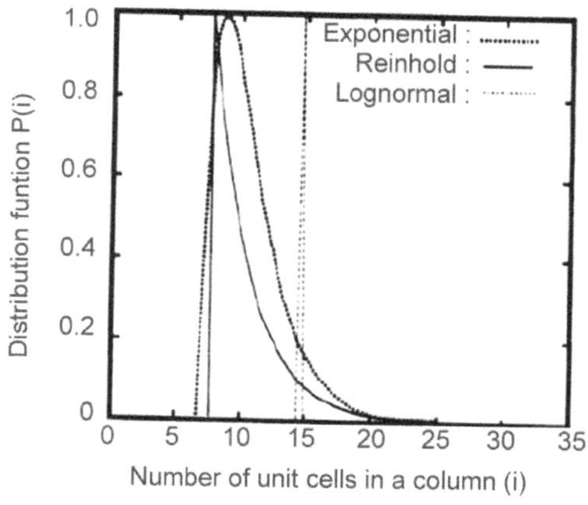

Figure 6.2 Crystal size distribution variation (eri fine silk fibre) along [201] the direction [15]

The reduction in crystal size value for the fine fibers as compared to coarse fibres is also supported by tensile properties like tensile strain and toughness [6]. The computed microstructural parameters have been used for computing the shape of coherent domains in terms of the shape of ellipsoid by taking the surface weighted crystal size value corresponding to (201) direction along X-axis and (100) direction along Y-axis for non-mulberry coarse silk fibres [15]. It is evident from Figure 6.3 that there are significant changes only in the periphery of the crystallite-shape ellipsoid. The shape could not be estimated in mulberry silk due to the lack of X-ray data.

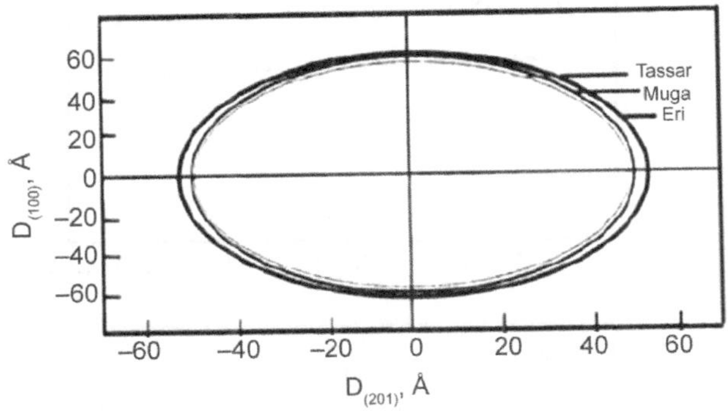

Figure 6.3 Crystallite-shape ellipsoid variation (non-mulberry coarse silk fibre varieties) [15]

Figure 6.4(a and b) depict the variation in tenacity with crystallite size for coarse and fine fibres of mulberry and non-mulberry silks. The tenacity for coarse and fine fibers of non-mulberry does not show significant change, the reason being the corresponding changes in the estimated microstructural parameters.

(a) Coarse silk fibre varieties

(b) Fine silk fibre varieties

The differences between mulberry and non-mulberry silk fibres arise due to the significant changes in the amino acid compositions, which in turn lead to a different molecular packing and crystalline morphology. Also, it has been reported that in mulberry, 29% of alanine, 45% glycine, and 12% of serine are present, whereas in non-mulberry silk fibres, 41% of alanine, 27% glycine and 11% of serine are present. Mulberry silk fibre has a crystal structure whose space group is $P2_1$ with β-pleated structure and fibre axis along b-axis whereas non-mulberry silk fibre belongs to $P2_12_12_1$ with a difference in β structure and a fibre axis along c-axis. These structural differences might lead to different crystalline regions and hence the wide angle X-ray patterns. It has been established that alanine is the material which concentrates in helical regions and also designated as helix-making, whereas glycine is found to be concentrated in non-local helical regions and referred to as helix-indifferent. These aspects have been quantified here in terms of microstructural parameters obtained by Fourier method, which has been appreciated by the round robin test.

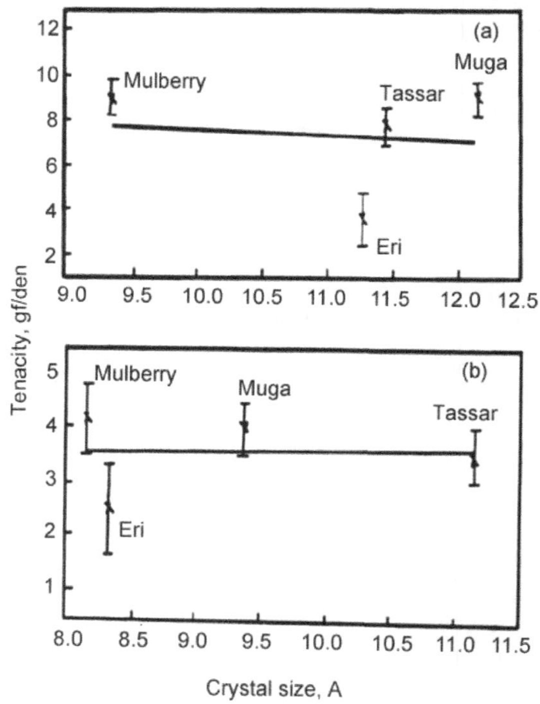

Figure 6.4 Tenacity variation with crystal size [15]

References

1. Okuyama K, Takanashi K, Nakajima Y, Hasegawa Y, Hirabayashi K and Nishi N, Journal of Sericultural Science, 57(1)(1988)23.
2. Freddi G, Gotoh Y, Mori T, Tsutsui I and Tsukuda M, Journal of Applied Polymer Science, 52(1994)775.
3. Gulrajani ML, Sen R and Chattopadhyay, A report on studies in reeling of tassar (Indian Institute of Technology, New Delhi) 1996.
4. Lizuka E, International journal of wild silk moth, 1(2)(1994)143.

5. Somashekarappa H, Annadurai V, Sangappa, Subramanya G, and Somasekhar R, Materials Letters, 53(2002)415.
6. Kushal Sen and Murugesh Babu K, Journal of Applied Polymer Science, 92(2) (2004)1095.
7. Kushal Sen and Murugesh Babu K, Journal of Applied Polymer Science, 92(2) (2004)1116.
8. Kushal Sen and Murugesh Babu K, Journal of Applied Polymer Science, 92(2) (2004)1080.
9. Rajkowha R, Gupta VB and Kothari VK, Journal of Applied Polymer Science, 77(1999)2418.
10. Hosemann R, Colloid Polymer Science, 260(1982)864.
11. Somasekharappa H, Somashekhar R, Vasudeva Singh and Ali SZ, Bulletin of Material Science, 22(1999)805.
12. Marsh RE, Corey RB, & Pauling L, Acta Crystal, 8(1955)710.
13. Okuyuma K, Somashekar R, Noguchi K, and Ichimura S, Biopolymers, 59(2001)310.
14. Conformation in fibrous proteins, edited by RB Fraser and TP Mactae (Academic Press, New York) 1989, 289.
15. Somashekarappa H, Mahesh SS, and Somashekar R, Indian Journal of Fibre and Textile Research, 30, 309.

7

Measurement of cohesion is silk yarn

Summary : There exists a good relation between normalized peeling force and cohesion strokes as measured from Duplan cohesion tester. With the increase in sericin content, the peeling force swiftly increases, more particularly at higher residual sericin content. The peeling force as measured on Instron is much more sensitive to the change in cohesion values particularly at low sericin content. Mulberry yarns are found to have greater cohesion than tussah yarns. The new method has good practical relevance in the silk industry in that the measurement of the silk yarn cohesion based on the peeling force is more accurate and reliable, and it could substitute the need of Duplan cohesion tester.

7.1 Introduction

The fibroin fibre is covered by the sericin protein with consecutive sticky layers which ensures the cohesion through sticking together of the silk filaments [1, 2]. A number of studies have been related to sericin from various types of cocoon shells [3–8]. In the silk reeling process, the cocoon filaments are bonded together. The extent of agglutination of such cocoon filaments bonding forming raw silk thread is considered crucial in silk yarn processing, which is very preferable aspect in silk yarn to withstand stress and strain in weaving. Obviously, higher cohesion lessens the end down rate during weaving, by way of the reduction of fraying and entanglements. Evaluation of silk yarn cohesion is not only important from the viewpoint of quality control, but it also helps to improve the processing performance of yarns. To improve silk yarn cohesion, some of the factors such as temperature, amount of reeling tension, sufficient croissure, and good drying of raw silk need to be carefully considered [9]. Cohesion of mulberry silk filaments was studied by Manna et al., using conventional cohesion tester [10]. Duplan cohesion tester has been used for evaluation of silk yarn cohesion. It works on the principle of applying strokes on a bed of silk threads held under tension till fraying appears (i.e., open places of 6 mm or more) at 10 different locations [11]. In general, machine is stopped after every 10 strokes and every single yarn is inspected very carefully to see if there are any fraying or open places having a dimension of 6 mm or more. If the number of these open places is less than 10, then another set of 10 strokes is applied and so on. The method has certain inherent shortcomings which may be read as follows. First of all, manual counting of

DOI: 10.1201/9781003735427-7

the number and size of open places is not only monotonous but also laborious too and hence may lead to erroneous results. Moreover, since the strokes are administered 10 at a time, invariably the number of frayed portions is more than 10, when the test is stopped. Hence, end point determination is ambiguous in nature. In addition, the yarn is considered to have been frayed at a particular location if the opening is 6 mm or more in length. In actual practice, the length is not measured. Hence the result relies more on individual judgement of the person performing the test and thus susceptible to be subjective. There seems to be a necessity for developing an alternate technique that is free not only from subjective errors, but also much more sensitive and reliable for scientific studies. As sericin is a binder and holds the filaments together, the force of cohesion between filaments could well be a determinant to measure cohesion. An attempt has been made to determine the peeling force necessary to separate out the filaments of various silk yarns.

7.2 The findings

The presence of residual sericin becomes the main source of cohesion. Other than sericin, the cohesion is also based on its nature and total available surface area of contacting mutual filaments. The number of filaments in yarn cross section, its fineness and yarn cross sectional shape determine the available surface area. For a given yarn count, higher filament fineness and more number of filaments would provide greater surface area of contact and cohesion. The actual area of contact will however depend upon cross sectional shape of filaments too. Assuming filaments to be randomly distributed in the yarn cross section, it can be said that the more its cross sectional shape deviates from circularity, the less would be the possible contact areas. This is so because filaments with non-circular cross section will hinder close packing [13]. Further, when filaments from a yarn are peeled out into two equal parts, group-wise filament separation takes place keeping mutual contact between filaments within a group intact. Therefore, all the contact points between filaments are not likely to break. In the present case, since residual sericin content, yarn and filament fineness are different for all the yarns used in this study, a corresponding difference in cohesion and peeling force values would be obvious. Therefore, a strict comparison between peeling force values of the yarns is not possible. The peeling force per unit surface area or peeling force per broken contact point for normalizing the peeling force is required, in order to make a valid comparison.

The surface area of a filament is proportional to the $(N)^{1/2}$, where N is the filament denier, considering circular cross section:

$$P_s \propto \frac{P}{n_b\sqrt{N_D}}$$

P_s – Peeling force per unit surface area

P – Absolute peeling force

n_b – Number of filaments

N_D – Denier of each filament

Hence, the normalized peeling force value dependent on the surface area(P_1) has been calculated by the using equation:

$$P_1 = \frac{P}{n_b\sqrt{N_D}}$$

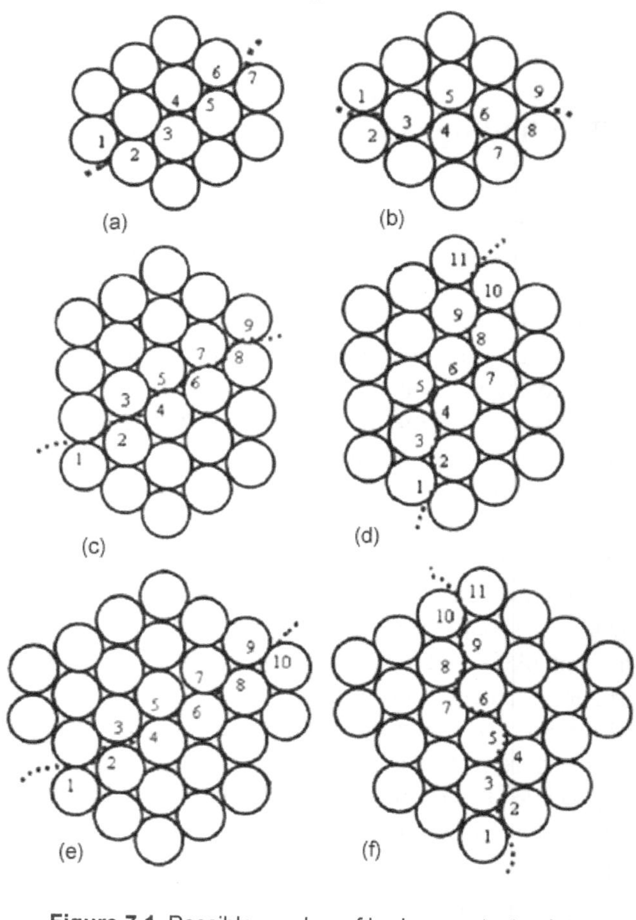

Figure 7.1 Possible number of broken contact points for hexagonal closed packed structure [13]

(a) 14 filaments in cross section with 7 broken contact points
(b) 14 filaments in cross section with 9 broken contact points
(c) 24 filaments in cross section with 9 broken contact points
(d) 24 filaments in cross section with 11 broken contact points
(e) 24 filaments in cross section with 9 broken contact points
(f) 28 filaments in cross section with 10 broken contact points
(g) 28 filaments in cross section with 11 broken contact points

For very highly packed structure, a hexagonal closed packed structure of filaments as postulated by Schwarz for circular fibres would be expected (Figure 7.1), where each circle represents a filament [12]. Since the separation may take place along any arbitrary line in the cross sectional plane of yarn as shown in Figure 7.1, the possible number of broken contact points may vary (Table 7.1). The peeling force based on the broken contact points (P_2) was calculated from the following equation

$$P_2 = \frac{P}{n_c}$$

where n_c is the number of broken contact points.

Table 7.1 Silk yarn specifications of various varieties [13]

Yarn type	Yarn size (den.)	Filaments/ C.S	Possible number of broken contact points
Tussah A	91	14	7, 9
Tussah B	52	24	9, 11
Tussah C	78	28	10, 11
Mulberry D (Multivoltine)	19	14	7, 9
Mulberry E (Multivoltine)	20	14	7, 9

Table 7.2 shows the cohesion data from Duplan cohesion tester and instron tensile tester. The results from Duplan cohesion tester demonstrate that the mulberry type E shows the highest value of cohesion followed by mulberry D and tussah type C. No cohesion is observed for type A and type B tussah yarns. The peeling force determined on Instron also shows a highest value for mulberry type E followed by tussah type C, mulberry type D and tussah types B and A. Peeling force therefore can be discriminated between tussah types A and B, which Duplan cohesion fails to do. Low peeling force values are observed type A and type B tussah yarns. A comparison in terms of normalized peeling force that the cohesion increases in the order: type

A < type B < type C < type D < type E. The order between type C and type D changes in comparison to the order obtained based on absolute peeling force and matches with that obtained on cohesion strokes.

Table 7.2 Peeling force and cohesion for different varieties of tussah and mulberry silk yarns [13]

Yarn type Duplan	Residual sericin (%)	Cohesion strokes	P, cN	P_1, $cN.den^{-1/2}$	P_2^2, cN
Tussah A	3.14	0.00	0.46	0.018	0.058
Tussah B	5.02	0.00	0.84	0.033	0.084
Tussah C	7.78	21.00	4.80	0.145	0.457
Mulberry D	20.70	31.00	4.22	0.364	0.527
Mulberry E	24.70	43.33	5.12	0.432	0.64

The grading in terms of normalized cohesion therefore remains identical to that obtained by cohesion strokes but becomes more sensitive. A high correlation coefficient between cohesion measured on Duplan and normalized peeling force based on surface area as well as number of broken contact points (Fig. 7.2) also substantiate this fact.

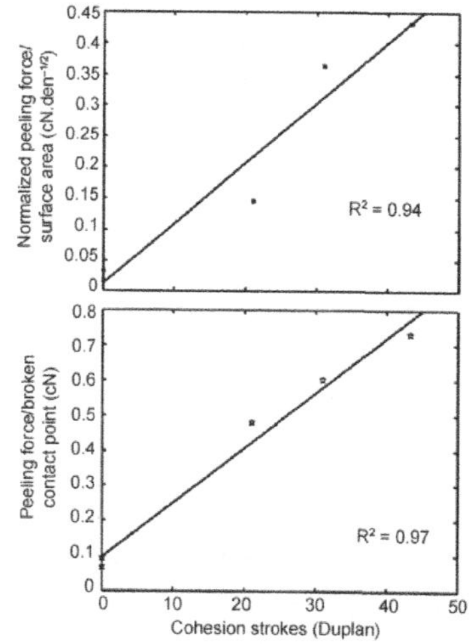

Figure 7.2 Correlation between cohesion strokes and peeling force for tussah and mulberry silk yarns [13].

Tussah yarns are normally less cohesive compared to mulberry yarns. It can also be seen (Table 7.2) that mulberry variety of type E which contains highest residual sericin also exhibits highest cohesion and peeling force [13].

A high level of sericin content and finer filament denier may be ascribed to this. Amongst tussah, type C shows highest value of cohesion and contains maximum amount of sericin as well. Besides it contains maximum number of filaments. Cohesion strokes are almost zero for residual sericin level of 5% or below. Peeling force however shows a reasonable value, thus indicating its sensitiveness to the low level of sericin. It appears therefore that the residual sericin and characteristics of tussah variety affect cohesion values. The residual sericin level in type C tussah yarn has been varied through various durations so as to determine the influence of only residual sericin on peeling force. The peeling force depending on the number of broken contact points increases with the increase in residual sericin content, initially slowly but becomes too rapid beyond 5% level of sericin (Figure 7.3).

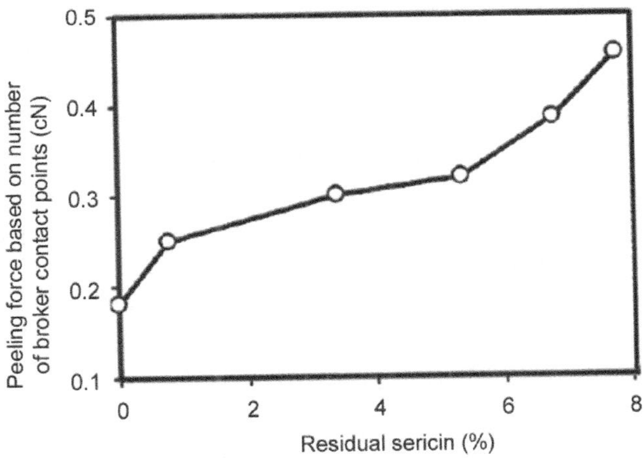

Figure 7.3 Influence of residual sericin on peeling force [13].

References

1. Zhang Y, Applications of natural silk protein sericin in biomaterials, Biotechnology advances, 20(2002)91–100.
2. Zhang Y, Tao M, Shen W, Zhou Y, Ding Y, Ma Y, and Zhou W, Immobilization of L-asparaginse on the microparticles of the natural silk sericin protein and its characters, Biomaterials, 25(17)(2004)3571.
3. Chattopadhyay R, Das S, Gulrajani ML, & Sen K, A study on the progressive change in characteristics of the bave (filament) along its length in mulberry and tasar cocoons, Sericologia, 37(2)(1997)263.
4. Pandey RK and Goel RK, Pro-anthocyanidins in the cocoon shell of 3 Antheria species, Sericologia, 30(4)(1990)527.

5. Das S, Chemical processing of tasar silk, Chemical processing of silk, edited by M.L. Gulrajani (Indian Institute of Technology, New Delhi), 1993, 46.
6. Tikoo BL and Goel RK, Oak tasar cocoon: Development of a simple cooking method, Indian silk, 25(9)(1987)55.
7. Kholmuminov AA, Bekmirzaeva GZ, Yuldashev MK, and Kabulov BD, Thin layer chromatography of the sericin, fatty waxes and fibroin of natural silk, Chemistry of Natural Compounds, 31(4)(1995)489.
8. Takasu Y, Yamada, and Tsubouchi K, Isolation of three main sericin components from the cocoon of the silk worm Bombyx mori, Bioscience, Biotechnology and Biochemistry, 66(12)(2002)175.
9. Lee Y, Silk reeling and testing manual, Raw silk reeling, FAO agricultural bulletin no. 136 (Food and Agriculture Organization of the United Nations, Rome), 1999, Chapter 6.
10. Manna SS, Sengupta D, and Rao GS, Measurement and comparison of cohesion in silk filaments in silk strands, Indian Journal of Textile Research, 14(1989)191.
11. Handbook of textile testing, ISI test method IS 2948 (Bureau of Indian Standards, New Delhi), 1964,190.
12. Schwarz ER, Certain aspects of yarn structure, Textile Research Journal, 21(3)(1951)125.
13. Das S, and Ghosh A, Evaluation of silk yarn cohesion based on peeling force, Indian Journal of Fibre and Textile Research, 34(2009)31.

8

Relating the microstructure and microrheological factors of different silk fibres

Summary: The silk varieties muga, eri and tassar have been compared for microstructure and microrheological parameters. Muga has greater mechanical strength than eri and tassar silk fibres at all the measured frequencies, a result that correlates with the calculations of the crystallite shape ellipses computed using the profile analysis of WAXS data recorded from these fibres. The shape ellipses do reveal information about the crystallite region of the fibres in muga silk variety that contributes to making these fibres stiffer than the other two varieties. This demonstrates the correlations between the techniques in the measurement of fibre stiffness. The investigations show that muga has greater stiffness than the other non-mulberry silk varieties as manifested in both forms studied. It is recommended in industrial production of silk-based materials which demand greater fibre stiffness.

8.1 Introduction

Silk gets spun into fibre by silkworm, spiders, scorpions, mites and flies [1, 2]. It is a semi-crystalline polymer comprising of two categories: mulberry (Bombyx mori) and non-mulberry silks (tassar, muga and eri), which are commercially important. Understanding the microstructural differences between them enables a comparative evaluation of their strength and properties. Investigations of macrostructure like cross-section measurements of silk fibres and densities of the Indian silk varieties have been reported by means of the analysis of silk fibroin amino acid contents [3]. The tensile stress–strain relation with XRD has been studied [4, 5]. It has been found that the stress relaxation is significantly greater in non-mulberry silk than in mulberry silk and that the differences among non-mulberry silk fibres are relatively small. The mechanical properties of regenerated silk fibroin (RSF) of polymer solutions of Bombyx mori silk using microrheological techniques have been studied earlier [6]. The unique mechanical properties of RSF make it especially interesting for many different applications, apart from its wide use in the textile industry, as mentioned earlier [7–9]. A recent study on the effect of shear field (spinning rate) and temperature on rheological properties of native silkworm and native spider dope has shown that both dopes behave

DOI: 10.1201/9781003735427-8

like polymer melts [10]. These findings provide a better understanding of silk formation from the silk worm, and have tremendous potential in enabling industrial production of silk that matches the qualities of natural silk. In the present work, wide angle X-ray scattering (WAXS) technique has been used to study the microcrystalline parameters such as crystal size (N) and lattice disorder (g in %) by employing Fourier method [11]. The study also reports the line profile analysis of Bragg reflections observed in X-ray patterns of three Indian silk fibres. Crystal imperfection parameters are computed with exponential distribution functions for each of the samples, after adopting robust refinement procedures. Crystallite shape ellipses were obtained by plotting the crystallite size along two orthogonal directions of the crystal for all the silk fibres. The areas of these ellipses are indicators of the differences in stiffness between the silk fibres. Video microscopic technique has been used to study the micro rheological properties of the RSF solutions. The position of an embedded tracer polystyrene bead is recorded in silk fibroin solution of the same concentration for all the three varieties in order to measure the storage and loss moduli and thus to characterize the viscoelastic properties of these media. The computed microrheological parameters from such an analysis are compared with microstructural parameters of the silk fibres. This study further reports the analysis of the relative stiffness of two different forms of the silk: one in crystalline form and the other in polymer solution form. The studies reveal a correlation in the relative strengths of the forms, for all the silk varieties. As an instance, the muga silk variety, which exhibits the largest crystallite shape ellipse area and thus higher stiffness, also shows in solution form the maximum viscoelastic modulus at all frequencies in investigations comparing all three varieties. The correlation studies between these different investigation methods have not been reported till time.

8.2 X-ray diffraction pattern

The X-ray diffractions of the polymer samples were recorded by special diffractometer having radiation of wavelength $\lambda = 1.506$ A. The scattered beam has been directed on a detector. The specifications used for the recordings were 40 kV, 30 mA. The polymer samples have been scanned in the 2□ range 12°100°. Figure 8.1 shows the X-ray diffraction patterns obtained for tassar, muga and eri silk fibres which are essentially silk II modification.

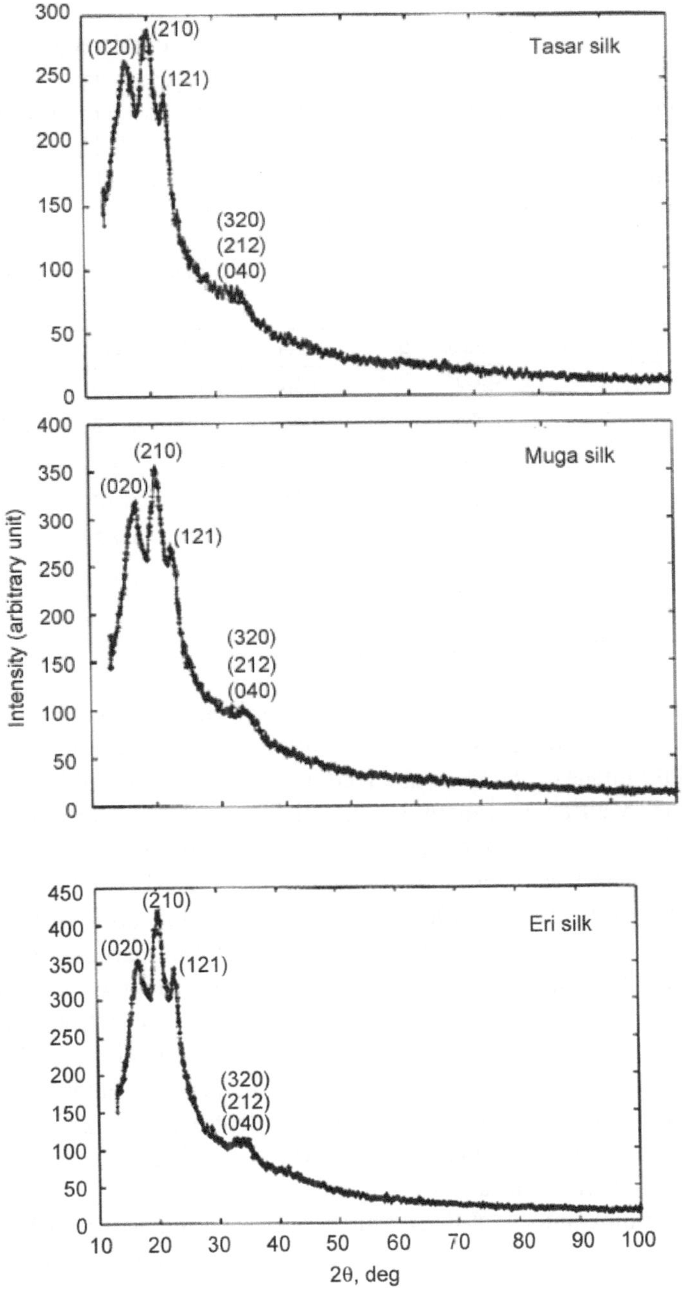

Figure 8.1 X-ray diffraction patterns of tassar, muga and eri silks

8.3 X-ray analysis

X-ray diffraction data have been used in a number of equations for the analysis, to simulate the intensity profile by varying the necessary parameters for obtaining a good fit with the experimental profile. Hence, a multi-dimensional algorithm SIMPLEX has been used for minimization [12]. The computed crystal imperfection parameters are given in Table 8.1 for exponential differentiation functions for each of the samples.

Table 8.1 Microstructural parameters of tassar, eri and muga silk fibres by using exponential functions [17].

Fibre	Profile no.	2□ deg.	g, %	delta	d_{hk1}, A	N	D_s, A
Tassar	1	16.62	0.5	0.06	5.33	2.85	15.19±0.91
	2	20.08	0.5	0.05	4.42	5.03	22.23± 1.11
Muga	1	17.04	0.3	0.05	5.20	5.47	28.44± 1.42
	2	20.08	0.1	0.05	4.42	5.10	22.54± 1.13
Eri	1	16.76	0.5	0.04	5.28	4.62	24.39± 0.97
	2	20.08	1.0	0.04	4.42	5.61	24.79± 0.99

8.4 Video microscopy

An open microscope has been used to record the Brownian motion of the beads in 1% RSF solutions of three silk varieties [13]. For every measurement, about 1000 images have been obtained at a rate of 10 fps, after focusing the microscope well above (about 30 μm) the cover slip so as to prevent bead surface hydrodynamic interactions. The on-screen magnification of the imaging system was determined by camera pixel density of 1280×1024 and optimal magnification of ×100. The on-screen magnification of the imaging system is a 63 nm pixel for our system. Through multiparticle tracking, 5–8 beads were tracked simultaneously for each field of view, and this helps in saving time while obtaining good statistics of the data from only a few movies. A feature-finding algorithm originally by Crocker and Grier in IDLVM is used for tracking the beads in non-deinterlaced fields of the movie. The tracking measurements are based on a centroid localization algorithm performed on sequential image frames, which can detect displacements of a bead with a precision of a few tenths of a nanometer [14, 15]. The selection of the algorithm and signal-noise ratio as well determines the final resolution achievable depends not only on the choice of the algorithm but also on signal-to-noise ratio of the raw data.

8.5 Findings of the investigations

A number of equations have been used to achieve the simulated profile with appropriate model parameters. The fit between the simulated and experimental profiles was approximately 50% (not shown here). Table 8.1 shows the computed microcrystallite parameters, like crystallite size (number of unit cells) (N), lattice strain (g in %), and the standard deviation. It is observed that the lattice strain in raw wild variety of silk fibre is small in muga and reasonably large in tassar fibre. Here, the standard deviation in all the cases for the microstructural parameters is taken as Delta (Table 8.1). A graphical part of the crystallite shape ellipse was obtained by taking the crystal size value corresponding to 2□ = 16.5° (020) direction along the X-axis and the other parameter corresponding to the 2□ = 20° (210) direction along the y-axis from the Figure 8.1 data. The angle between (020) and (210) planes is around 90°. Figure 8.2 depicts the crystallite shape ellipse for the types of silk

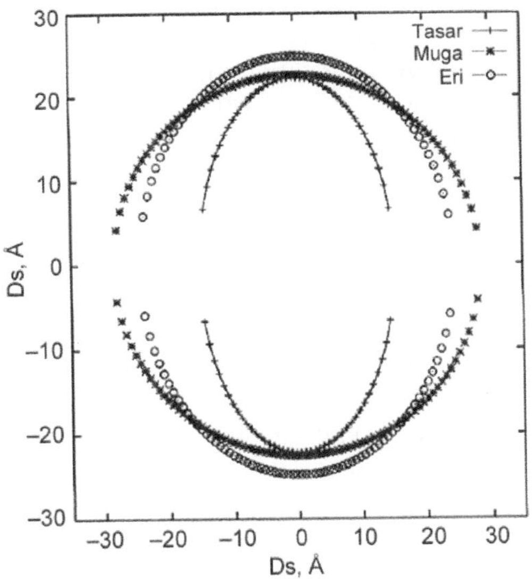

Figure 8.2 Difference of crystallite shape in raw silk wild silk varieties [17]

The strength of silk fibre is normally proportional to crystalline area which is equal to ellipse area determined by microstructural parameters. It is evident that the crystallite shape ellipse in muga is greater compared to eri and tassar, while the value of the crystallite shape area of eri lies in between muga and tassar. Sen and Murugesh found that muga exhibits the highest density, followed by tassar and eri among the non-mulberry silk varieties.

This suggests that the degree of crystallinity and crystallite orientation are more in muga than as compared to other varieties. The WAXS studies suggest that the eri have highest intensity (>400; Fig. 8.1) for 110 planes than muga (>350) and tassar (>300) silk, indicating the high degree of crystallinity for eri silk, although its absolute measurement is difficult. The video microscopy set up for rheological studies has been calibrated using Newtonian liquids such as pure water and pure glycerol. The details of this calibration procedure can be found elsewhere [6]. Briefly, the measured viscosities for water and glycerol are found to be in agreement with the standard values at 22°C [17]. This microrheological technique is suitable for characterizing mechanical properties of liquids. With this set up, one can probe materials having shear modulii values lying between 5×10^{-4} Pa and 1 Pa [6]. For these reasons, concentrations of silk sample are chosen so as to lie within the measurement range of set up. In video microscopy, the static error ε_2 in tracking the embedded beads was estimated by analyzing the positions of 2a = 0.989 beads stuck to the cover slip in 1000 frames maintaining nearly the same environment used for the analysis of RSF samples. This gives the value of spatial resolution ε_2 = 18 nm from the MSD via the relationship $\varepsilon_2 = \sqrt{(\Delta r^2(\tau)/2}$ at short lag times for which statistical accuracy is the best [16]. At longer lag times (above 60 s), the static error for stuck beads in our set up, using feature tracking, is increased to a value of 33 mm due to the statistical accuracy in obtaining MSD with only few data points. It was also found that at even longer lag times this static error further increases (though marginal) with lag time. The reason for this monotonous increase at high lag times is yet to be understood. Further, dynamic error also contributes to spatial error during particle tracking in microrheological studies. It has been found that the spatial error follows the same temporal distribution as the pixel intensity noise in the movie. The dynamic error depends on the exposure time σ of the camera [16]. With these spatial errors could be given as $(\Delta r^2(\tau, \sigma))_{measured} = 2D(\tau - \sigma/3) + 2\varepsilon_2^2$ where $D = k_BT/6\pi\square a$ is the beads diffusion coefficient in a medium of viscosity □. By using σ value to 1/10th of lag time (i.e., 0.01s), it has been ensured that the contribution of dynamic error to MSD is minimum. Super diffusion in the MSD values at very short lag times (between 0.1 s to 0.4 s) has been observed due to the dynamic error in the set up. For this reason the data up to 0.4s has been removed from the MSD values. From the measured value of ε_2 for the entire range, one can remove static error by using equation (10), to obtain static error-free MSD ($((\Delta r^2(\tau, \sigma))_{sfree}$) data by modeling it as

$$((\Delta r^2(\tau, \sigma))_{sfree} = ((\Delta r^2(\tau, \sigma))_{measured} - 2\varepsilon_2^2$$

Though the dynamic error still persists in the data, this method of removing static error is valid whenever one estimates average properties, such as MSD

[16]. This calibrated set up was used for recording the Brownian motions of beads in all the three varieties of silk at a room temperature of 23.5±0.5°C and then analyzed offline. The ensemble averaged MSD of several (80–100) beads in 1% RSF of eri, muga and tassar as a function of lag time was calculated using Eq. (8). The static error corrected MSD data using above equation (Fig. 8.3a). All the curves in this plot have varying slope varying between 0 and 1, a characteristic of viscoelastic materials. At small lag times, the material shows solid-like behavior due to the presence of the polymer network of RSF. At higher lag times, RSF sample shows liquid-like behavior, as the bead diffuses freely in the sample [17]. The MSD of embedded beads in tassar RSF is greatest and that of muga is lowest at all frequencies, while the MSD of embedded beads in eri lies in the middle. Note that the MSD of beads in pure solvent is much higher than that in the sample for the entire frequency range [6]. The shear moduli of the three samples were calculated using Eq. (9) (Fig. 8.3b).

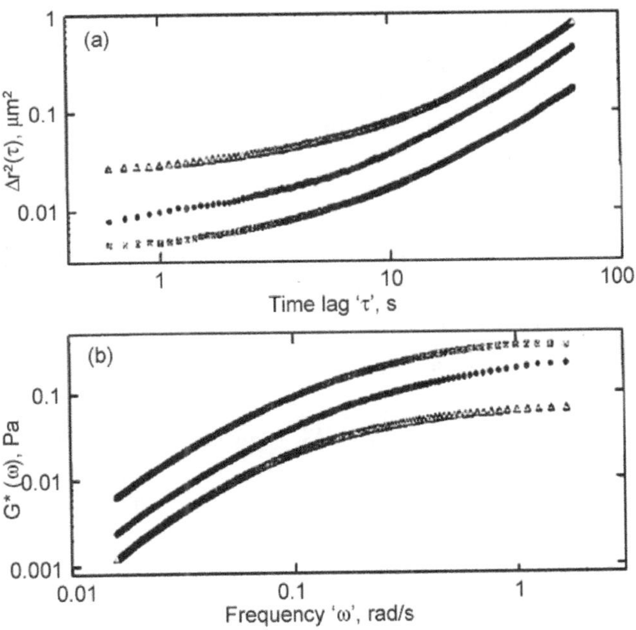

Figure 8.3 (a) Ensembled mean square deviation of beads of 1% RSF of tassar (Δ), eri (.) and muga (x) as a function of lag time (b) Corresponding shear moduli of three samples [17].

The shear modulus of tassar RSF is lower and that of muga and eri is higher at all frequencies. This demonstrates that muga and eri are stiffer than tassar. In addition, higher mechanical strength suggests that muga has higher molecular weight than other RSF samples. Intrinsic viscosities of 0.76, 0.71 and 0.69 dL/g have been reported respectively for muga, tassar and eri, and this trend is borne out in our measurements of shear moduli of the varieties, except in the relative values of viscosities of eri and tassar [3]. The difference may be due to difference in the varieties of silk fibres used in this investigation compared to that reported earlier [3]. The relative strengths of the fibres measured in terms of their shear moduli using microrheological techniques, and the fibre stiffness characterized through crystallite shape ellipses computed from WAXS data, reveal an interesting correlation between fibre strength and corresponding shear moduli of the same fibre in solution form. The study shows that the silk polymer chain that has the highest (lowest) relative stiffness measured by WAXS analysis also gives rise to the highest (lowest) shear modulus in solution form at all frequencies. The correlation appears to be based on the possibility that the entangled network of the fibres having higher stiffness is more resistant to the tracer bead's movement when it is embedded within a mesh formed by the entangled structure leading to polymer solution with higher shear modulus. In order to properly understand this correlation and its possible limits, further studies are required, which would enable for further work on the microstructure-rheological property.

References

1. Kaplan DL, Adams WW, Farmer B & Viney C, Silk polymers: Materials science and biotechnology, ACS symposium series, 544(1994)2.
2. Marsh RE, Corey RB, and Pauling L, Biocem Biophys Acta, 16(1995)1.
3. Kusal Sen and Murugesh Babu K, Journal of Applied Polymer Science, 92(2) (2004)1080.
4. Rajkhowa R, Gupta VB, and Kothari VK, Journal of Applied Polymer Science, 77(11) (2000)2418.
5. Kothari VK, Rajkhowa R, and Gupta VB, Journal of Applied Polymer Science, 82(5) (2001)1147.
6. Raghu A, Somasekhar R, and Sarath Anantamurthy, Journal of Polymer Science/part B/polymerphysics, 45(2007)2555.
7. Xin C, Knight DP, Shao Z, and Volrath F, Polymer, 42(2001)9969.
8. Kaplan DL, Fossey S, and Mello CM, MRS Bulletin, 17(1992)41.
9. Li M, Wu Z, Zhang C, Lu S, Yan H, Huang D, and Ye H, Journal of Applied Polymer Science, 79(12)(2001)2192.
10. Holland C, Terry AE, Porter D and Volrath F, Nature Material, 5(2006)870.
11. Warren BE, X-ray diffraction (Addision Wesley, New York) 1969.

12. Numerical Recipes, edited by W Press, BP Flannery, S. Trulolsky, and WT Vetterling (Cambridge University Press, UK) 1986.

13. Raghu A, and Sharath Ananthamurthy, Pramana Journal of Physics, 65(2005)69.

14. Crocker JC, and Grier DG, Journal of Colloid Interface Science, 179(1996)298.

15. Breedveld V, and Pine D, Journal of Material Science, 38(2003)4461.

16. Thierry Savin, and Patrik S Doyle, Biophysics Journal, 88(2005)623.

17. Divakara S, Somashekar R, Raghu A, Yogesha, Sharath Ananthamurthy, and Subrata Roy, Indian Journal of Fibre and Textile Research, 34(2009)168.

9

New method of evaluation of low-stress mechanical properties of silk fabrics

Summary : The simple conventional method of determining low-stress mechanical properties of fabrics has been refined so as to render it suitable for finished silk fabrics. The tensile stress–strain curves have been used to obtain the bending rigidity of the fabrics and bias extension for shear rigidity. The correlation between the parameters obtained by the modified method and that obtained by the Kawabata evaluation test method is found to be good for a series of finished silk fabrics.

9.1 Introduction

In order to understand handle and tailorability of fabrics, the low-stress mechanical properties become the base for understanding. There are many noteworthy contributions that enable good understanding of low-stress mechanical properties of fabrics. A new technique has been evolved for measuring the low-stress mechanical properties of woven fabric [1]. The principle is based on the fact that when a concentrated load is applied on the boundary of the very large plane, it causes buckling, which is the basic mode of bending, shearing and tensile deformations. They found that the tensile, bending and shear properties of woven fabrics can be obtained from the slopes and strains of the curves. They also demonstrated that the slopes in the post-buckling region provide the shear rigidity. They reported the correlation between their method and Kawabata method of measuring low-stress mechanical properties. Yazdi sufficiently used the techniques for studying a large number of fabrics for their mechanical properties [2]. This methodology is interesting as it does away with the use of Kawabata instruments. Some simple methods of measuring have also been developed notably by Amirbayat and Alagha [3]. There are two approaches to measure low-stress mechanical properties, namely,

(a) By extracting the sample through a nozzle or ring using a force measuring device, and

(b) By utilizing Kawabata evaluating system and FAST [4].

The high cost, involved in testing on these systems, precludes their use in a number of laboratories. Hence, some indirect methods of measuring low-stress mechanical properties have been developed, and one such method is

DOI: 10.1201/9781003735427-9

bias testing of fabrics on Instron developed by Amirbayat and Alagha [3]. Further, it was found that the methodology developed by Yazdi and Amirbayat is inapplicable to silk fabrics in view of the complete absence of yield point in the tensile curves of the degummed sample. There is a need to develop an alternate solution to determine bending and shear rigidity of the silk fabric so as to circumvent this problem. Study has thus been done to modify the conventional technique to render it suitable for measuring the low-stress mechanical properties of raw and degummed silk fabrics.

9.2 Technical details

Mulberry bivoltine raw silk fabric has been used for the investigations. It has been subjected to the following treatments:

(a) Degumming
(b) Dyeing – Acid, metal complex, and reactive at three different concentrations
(c) Printing – Direct and discharge styles on dyed fabrics
(d) Finishing – Calendering, decatising, and stain guard finishes.

Equipments used – Manually operated tensile tester, Instron tensile tester, and KES–F shear tester.

The method suggested by Amir Bayat and Alagha for evaluating bending, tensile and shear properties from the bias testing of fabric was found to be inapplicable for silk fabrics in view of the complete absence of yield point in the tensile loading and recovery curves [6].

9.3 Findings of the investigations

The load extension curves for three kinds of polyester, raw silk, and degummed silk fabrics are shown in Figure 9.1. There is no yield point in degummed silk fabric although it is present in raw silk and other fabrics. The shape of the curve is concave confirming that the degummed silk fabric does not show any yield point [6]. In order to circumvent this problem, an alternate solution to determine bending and shear rigidities has been introduced.

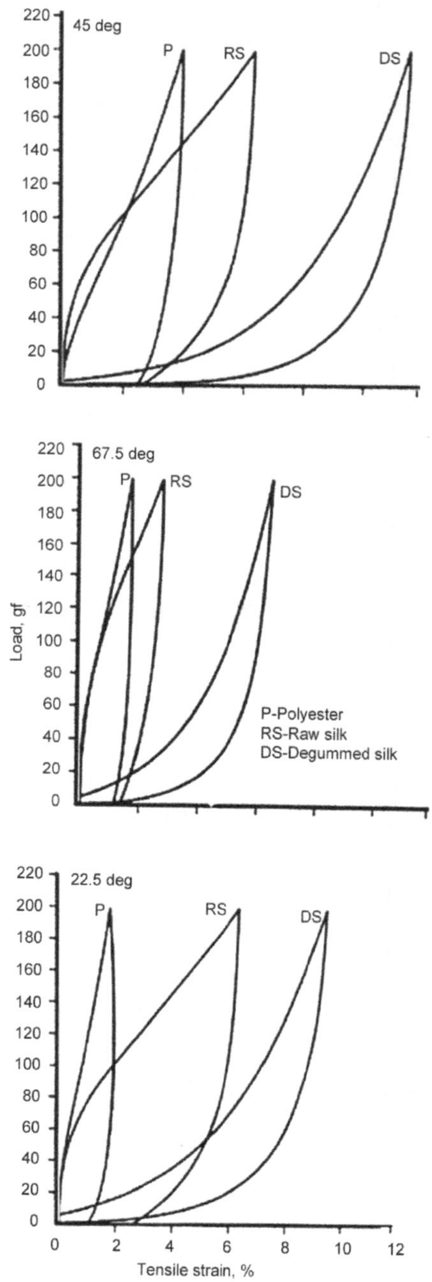

Figure 9.1 Curves depicting load elongation for polyester, raw silk, and degummed silk at different directions [6].

The values of slope determined from the tensile curves at 22.5°, 45°, and 67.5° have been compared with bending rigidities values of KES-F and correlated. The correlation between the Young's Modulus determined from the bias extension curve at 22.5° and bending rigidity obtained from KES-F is found to be 0.916, and the correlation between the Young's Modulus determined from the bias extension curve at 45° with bending rigidity of KES-F is 0.955. Since the Young's Modulus is determined from the bias testing of fabrics particularly at 45°, it is suggested that this method should be followed as a routine quality control test instrument [6]. Thus some modification in the procedure laid by Amirbayat and Alagha is required for the determination of bending properties of silk fabrics. These fabrics are found to be unique in the sense that by subjecting to degumming and subsequent finishing treatment, the shape of the tensile curve becomes concave (Figure 9.1). Shear rigidity values obtained by the bias extension at 45° using Instron were observed and compared with that measured by KES-F. These values along with the shear rigidity values obtained by manual method using simple instrument are found to be highly correlated. The correlation between Kawabata, Instron and manual methods is found to be excellent. Hence, it can be concluded that a simple method of determining the shear rigidity by bias extension can be used as the manual method and the method merits consideration in view of its very good correlation with Instron. The results obtained by Instron and KES-F are found to be correlated and that obtained by bias test and KES-F are well correlated. The results also show that the finishing treatments calendaring, decatising, and stain guard, also improve the total hand values of the silk fabrics. Calendering treatment has led to a substantial increase in elongation; decatising process has led to an increase in tensile properties; and stain guard finish has led to the lowest coefficient of friction and shows a reduction in tensile properties. The method developed by Yazdi for determining tensile, bending and shearing properties by bias cannot be applied for silk fabrics [1, 2]. Hence, this method has been modified which involves testing of the fabrics in bias directions and it is found that all the three parameters viz., tensile, shear and bending, can be obtained. The correlation between the fabric extensibility determined by Instron is high with the fabric extensibility determined by Kawabata tensile tester. Further, the shear rigidity determined using the modified method is highly correlated with that obtained using the Kawabata tensile tester. The bending rigidity determined by the new method and that determined by the Kawabata bending tester bear a very good correlation. The new method developed has industrial significance in that it is cost effective, less time consuming and suitable for a range of silk fabrics.

References

1. Alamdar Yazdi A, and Amir Bayat, Evaluation of the basic low stress mechanical properties, International Journal of Clothing Science and Technology, 12(2000)311–330.
2. Alamdar Yazdi A, A new method to evaluate low stress shearing behavior of woven fabrics, Indian Journal of Fibres and Textile Research, 29(2004), 333.
3. Amirbayat J, and Alagha MJ, A new approach to fabric assessment, International Journal of Clothing Science and Technology, 7(1995)46.
4. El Behety HM, Effect of mechanical and physical properties on fabric hand (Woodhead Publishing Limited, Cambridge, England and CRC Press LLC) 2005.
5. Kilby WF, Shear properties in relation to fabric hand, Textile Research Journal, 31(1961).
6. Radhalakshmi YC, Somashekar TH, and Subramaniam V, Suitability of modified method for evaluating low stress mechanical properties of silk fabrics, Indian Journal of Fibres and Textile Research, (34)(2009), 283.

10

Kinetics of adsorption of natural dye on silk: Effect of surface modification

Summary : The effect of plasma-treated silk surfaces on the adsorption kinetics and thermodynamics of lac dyeing on silk has been studied. The use of oxygen and Ar plasma for treatment of silk surfaces had a major influence on the conformation change from the amide-II random coil to the amide-I b-sheet structure. The silk surfaces were notably roughened by the oxygen, and Ar plasma treatments and surface roughness was found to increase with increasing plasma treatment time. Under the controlled initial dye concentration and the MLR, the optimum pH of dye solution was found to be 3.0. The kinetic mechanism of adsorption follows the pseudo-first-order equation for the initial stage of adsorption. The adsorption capacity showed good improvement in the case of plasma-pretreated samples, and the best results have been seen for lac dyeing of silk (Ar treated).

10.1 Introduction

Natural dyes are generally attractive, soft and exist in multi-hued shades. The dyeing of textile fibres with natural dyes has gained interest due to their compatibility with the environment and also their generally lower toxicity and allergic reaction. Lac dye is a popular reddish dyestuff obtained from a secretion of the insect Coccus laccae [1]. It finds applications in coloring food, as a cosmetics ingredient, oil painting and dyeing textiles such as silk, cotton and wool. Lac dye is composed mainly of two major groups of substance: four closely related laccaic acids, A, B, C and E, all of which are water-soluble pigments producing a red color, and erythrolaccin, which dissolves in alcohol producing a pale-yellow color [2–4]. The chemical structures of laccaic acids A, B, C and E are shown in Figure 10.1 (I).

Lac dye is widely used as a red dyestuff for cotton and silk dyeing, despite its restricted use on cotton or silk dyeing as it is not readily adsorbed by cotton or silk. Hence, the fastness properties and reproducibility to give consistency in production are still the problems to be solved. In general, the routes to solve these problems consist of fundamental physical studies on the dyeing process [5, 6] and chemical modifications to improve the lac dye adsorption on the fibers [7, 8]. For the latter route, various types of mordants were used as surface modifiers to promote the binding of dyes to fabric by forming a chemical bridge from dye to fiber, enhancing the staining ability of a dye

DOI: 10.1201/9781003735427-10

(I)

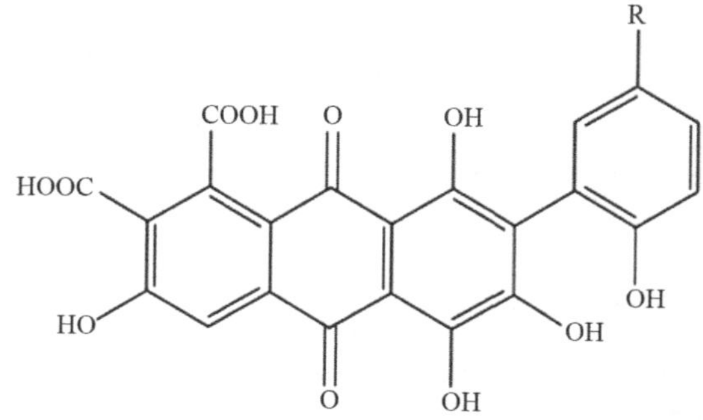

Laccaic acid	R	MW
A	$CH_2CH_2NOCOCH_3$	537
B	CH_2CH_2OH	496
C	$CH_2CH(NH_2)COOH$	539
E	$CH_2CH_2NH_2$	495

(II)

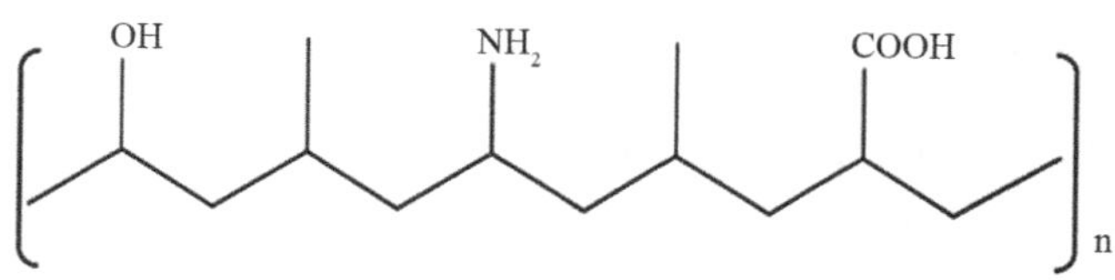

Figure 10.1 Chemical structures of laccaic acids (I) and silk fibroin (II).

along with increasing its fastness properties. However, most of the mordants used are toxic and have serious detrimental effects on the environment. Therefore, an environment-friendly method of surface modification is required. Low-temperature plasma treatment is a technique for modifying the fiber's surface to improve surface wettability, shrink resistance, interfacial adhesion, hydrophilicity, hydrophobicity and dyeing properties [9–16]. The advantages of this technique are that the plasma modification is only confined to the surface of the materials without interfering with their bulk properties, and the plasma process itself is environment-friendly, involving no chemicals. There are a few limitations for commercially utilizing this plasma process, including technical problems during system scale-up for industrial use. Moreover, the lifetime of treated samples still cannot meet the requirements of the textile industry on permanence against washing, light, perspiration and so on. Recently, the atmospheric-pressure plasma treatment has been developed and widely applied to textiles to reduce the drawbacks of the low-pressure plasma system [17, 18]. However, although many advantages are gained from the atmospheric-pressure plasma, the low-pressure plasma is still useful and convenient for surface treatment on the laboratory scale. In order to improve the adsorption capacity of lac dye on silk, attempt has been made to modify the silk surfaces using low-pressure plasma treatments on the laboratory scale. The influence of silk surface modification by oxygen and Ar plasma treatments on adsorption kinetics of lac dyeing on silk has been studied. The adsorption process kinetics has been compared for untreated and the pretreated silks containing dyeing systems.

10.2 ATR-FTIR characterization

In the surface analysis of textile materials, the ATR-FTIR spectroscopy is found to be well suited owing to its convenient, fast and non-destructive testing technique. Influence of oxygen and Ar plasma treatment duration on the structure of silks is shown by ATR-FTIR spectra. The different amide bands appeared in the range of 1700–600 cm^{-1}, which are characteristic of polypeptides and proteins [19].

For oxygen or Ar plasma-treated samples at each treatment time, the peak at 1540 cm^{-1} corresponding to the amide-random coil band is mostly disappeared. The change indicates a structural transition of the random coil conformation after plasma treatment of silk surfaces. At the same time, the amide-I b-sheet band (1628 cm) of silk seems to change with the plasma treatments. These findings are similar to those of silks pretreated with oxygen plasma reported by Chen et al. [20]. They reported that the conformation change from the amide-II random coil to the amide-I b-sheet structure of silk under oxygen plasma treatment arises from the oxidation and deoxidation reactions during and after treatment, respectively. Furthermore, the macromolecular group partly decomposed and reorganized, leading to the increase in the b-sheet structure. The pretreatment of silk surfaces using Ar plasma is found to mainly influence the conformation change from random coil to a b-sheet structure similar to that using oxygen plasma.

10.2.1 Surface morphology of silk

Figure 10.2 depicts the influence of oxygen and Ar plasma treatments during each plasma treatment on the surface morphology of silk as viewed by SEM.

In the case of untreated sample (Figure 10.2a) the smooth surface of the silk fiber is clearly observed, whereas the silk fibers were notably roughened by the oxygen and Ar plasma treatments (Figures 10.2b–e).

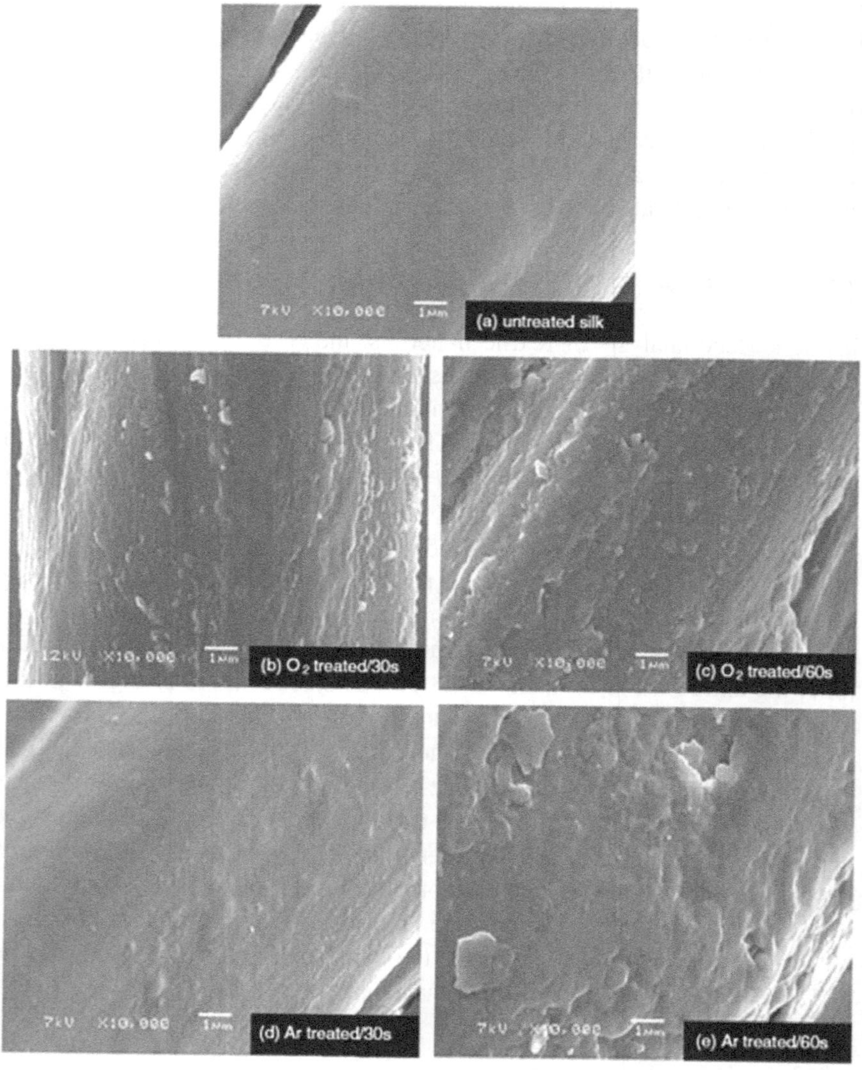

Figure 10.2 SEM photographs of (a) Untreated silk (b) Oxygen plasma-treated silk for 30s (c) Oxygen plasma-treated silk for 60s (d) Ar plasma-treated silks for 30s (e) Ar plasma-treated silks for 60s

Although the surface roughness of the silk fiber increases with increasing plasma treatment time, the surface morphology of silks, exposed to the plasma treatment greater than 60 s, are not shown here. Note that at the same plasma treatment time, the Ar plasma-treated silk shows relatively lower surface roughness than the oxygen plasma-treated one. Typically, oxygen plasma modifies the surface by grafting hydroxyl, carbonyl and carboxylate groups [19]. But argon plasma is known to be more efficient in introducing oxygen-containing groups with relatively low surface roughness than oxygen plasma, owing to the chemical etching nature of oxygen [21].

10.2.2 Influence of pH on lac dyeability

The pH of the dye solution is one of the important factors that influences the adsorption of lac dye onto silk. NH_2 groups of laccaic acids C and E and –NH– group of laccaic acid A interact with –OH groups of the silk surface, and –OH groups of all laccaic acids can interact with NH_2 groups of the silk surface. The dye solution pH was varied between 2.0 and 5.0 by adjustment with glacial acetic acid. The highest absorption capacity has been seen at pH of 3, depending on the amount of dye adsorbed per gram silk at equilibrium.

The highest dyeability of lac dye in such pH arises from an increase in the protonation of the amino group (NH_2) of amino acid in silk protein, whereas the carboxylic groups in the side chain are unionized at lower pH [5]. Also, at pH value 3, –NH– or NH_2 groups of lac are protonated. This observation indicates that under acidic conditions, there is electrostatic interaction between $-NH_{2+}$ and NH_{3-} groups of lac with –OH groups of silk surface and electrostatic interaction between –OH groups of lac with NH_{3+} groups of the silk surface. The reduction in pH value appears to increase the positive electric charge on the silk surface, and due to the repulsion between lac dye molecules and the silk surface, this change leads to reduction in qe value. Hence, the dye solution pH in all trials of the investigation has been fixed at 3.0.

10.2.3 Influence of plasma treatment duration on silk surface and dyeing temperature on lac dyeability

Influence of plasma treatment time

The conditions of the plasma technique play a crucial role to control the reaction at the surface and, thus lead to the desirable modified surface. One of the plasma treatment factors influencing dyeability is the optimum treatment duration. The adsorption capacity of lac dye onto the silk modified using oxygen and Ar plasma treatments at different plasma treatment times was investigated. The dyeing temperatures were set at 30, 60 and 90°C. It is obviously seen that the dyeability of the lac dye onto silk depends on the types of plasma gas, plasma treatment time and dyeing temperature. By comparing the qe value with that of the untreated silks, the qe value of the oxygen- and Ar-treated silks containing dyeing systems sharply increases with plasma treatment for 10 and 30 s. Particularly at dyeing temperatures of 60 and 90°C, the qe values of the treated silk systems under plasma treatment for 10 and 30 s are higher by about two times than that of the untreated silk. After being exposed to plasma treatments for longer times, the qe values of the plasma-treated silks gradually drop. At the same treatment time and dyeing temperature, the qe values of the Ar plasma-treated silk are mostly somewhat higher than those of the oxygen plasma-treated sample. From the results, the conclusion could be drawn that lac dyeing onto silk would be favorable for the pretreated silk surfaces using plasma treatment time for 30 s. Hence, the pretreated silk samples used for adsorption kinetic studies in the study have been modified using plasma treatments for 30 s.

Influence of dyeing temperature

In order to investigate the effect of dyeing temperature on the lac dyeability, the control of driving force due to the concentration gradient is essential. The concentration of lac dye solution has been fixed at 500 mg/l with the materials-to-liquor ratio (MLR) of 1:100 and pH 3.0. It is seen that for all samples, the amount of dye absorbed per gram of silk at any dyeing time (qt) increases sharply first, gradually increases and then roughly keeps constant after dyeing for 60 min. Before approaching the dyeing equilibrium, dyeing using a higher temperature results in a higher initial adsorption rate, as seen from the higher initial slope of the curves.

These findings show that the increase in dyeing temperature leads to the mobility of large dye ions and consequently the rates of adsorption improve. These results are in accordance with those observed for lac dyeing on untreated silk and cotton [5, 6]. However, the qt for all samples decreases with increasing temperature at equilibrium state, suggesting an exothermic adsorption process. By comparing the effect of plasma treatments on the silk surface, on the lac dye ability, the qe values of all dyeing systems were examined. The respective qe values for the dyeing systems containing the untreated oxygen and Ar plasma-treated silks were found to be 26.5, 42.5 and 46.8 mg/g silk at 30°C; 20.3, 37.1 and 43.0 mg/g silk at 60°C; and 18.4, 34.9 and 40.6 mg/g silk at 90°C. It is obviously seen that at the same dyeing temperature, the qe values of the plasma-treated silks are much higher than those of the untreated sample. The obtained results indicate that the adsorption capacity of silk is much improved with pretreatment of the silk surface using plasma techniques. It is known that oxygen plasma typically modifies the surface by grafting hydroxyl, carbonyl and carboxylate groups, and argon plasma treatment has a high efficiency in introducing oxygen-containing groups in the treated samples [19, 21]. Hence, increase in oxygen-containing groups of the silk-treated surface increases electrostatic interaction between the lac and the treated silk surface along with a change in the conformation of the silk structure. It improves lac adsorption on the surface of the treated silk over the untreated silk. In the case of pretreated silks treated with oxygen and Ar plasma, the qe value increases progressively implying that the highest adsorption capacity is achieved.

10.2.4 Investigations on Kinetic of adsorption

Pseudo-first and second-order kinetic models have been used to study the adsorption kinetics of lac dye on silks to analyse the experimental data. Lagergren equation has been used for the analysis of pseudo-first and second-order kinetics of adsorption [5, 6, 23, 24]. The plot of t = qe versus t would be a linear relationship, if pseudo-second-order kinetics are applicable. The adsorption capacity increases sharply first at 0–10 min. After that, the adsorption rate decreases and reaches equilibrium at about t = 60 min. The pseudo-first and pseudo-second-order equations were applied for adsorption in the time range of 0–10 min. The results show that the pseudo-first-order plot shows a good linear relation, whereas the pseudo-second-order equation does not fit well for this region. Hence, the results of the pseudo-first-order kinetic calculation have only been obtained. These results agree well with those of the adsorption kinetics of lac dyeing on cotton and the equilibrium and kinetic modeling of reactive dye on cross-linked chitosan bead [23]. It is also found that the pseudo-first-order kinetic equation of Lagergren does not fit well for the whole range of dyeing time and is applicable over only the initial stage of adsorption.

References

1. Moeyes M. Natural dyeing in Thailand. Bangkok: White Lotus, 1993.
2. Burkinshaw SM and Gotsopoulos A. Pre-treatment of cotton to enhance its dyeability - I. Sulphur dyes. Dyes Pigm 1996; 32: 209–228.
3. Oka H and Ito Y. Separation of lac dye components by high-speed counter-current chromatography. J Chromatogr A 1998; 813: 71–77.
4. Stana-Kleinscheck S and Ribitsch V. Electrokinetic properties of processed cellulose fibers. Colloids Surf 1998; 140: 127–138.
5. Chairat M, Rattanaphani S, Bremner JB, et al. An adsorption and kinetic study of lac dyeing on silk. Dyes Pigm 2005; 64: 231–241.

6. Chairat M, Rattanaphani S, Bremner JB, et al. Adsorption kinetic study of lac dyeing on cotton. Dyes Pigm 2008; 76: 435–439.
7. Janhom S, Griffiths P, Watanesk R, et al. Enhancement of lac dye adsorption on cotton fibres by poly(ethyleneimine). Dyes Pigm 2004; 63: 231–237.
8. Janhom S, Watanesk R, Watanesk S, et al. Comparative study of lac dye adsorption on cotton fibre surface modified by synthetic and natural polymers. Dyes Pigm 2006; 71: 188–193.
9. Molina R, Javancic P, Jocic D, et al. Surface characterization of keratin fibres treated by water vapour plasma. Surf Interface Anal 2003; 35: 128–135.
10. Jin JC and Dai JJ. A study of wool dyeing with nitrogen plasma treated dyestuff. J Textil Res 2002; 23: 9–10.
11. Moon SI and Jang J. Factors affecting the interfacial adhesion of ultrahigh-modulus polyethylene fiber vinylester composites using gas plasma treatment. J Mater Sci 1998; 33: 3419–3425.
12. Riccardi C, Barni R, Selli E, et al. Surface modification of poly(ethylene terephthalate) fibers induced by radio frequency air plasma treatment. Appl Surf Sci 2003; 211: 386–397.
13. Chaivan P, Pasaja N, Boonyawan D, et al. Low-temperature plasma treatment for hydrophobicity improvement of silk. Surf Coating Technol 2005; 193: 356–360.
14. Vohrer U, Muller M and Oehr C. Glow-discharge treatment for the modification of textiles. Surf Coating Technol 1998; 98: 1128–1131.
15. Kin MS and Kang TJ. Dimensional and surface properties of plasma and silicone treated wool fabric. Textil Res J 2002; 72: 113–120.
16. Chen YY, Zou LY and Sheng J. Influence of oxygen plasma treatment upon mulberry silk microstructure. J Textil Res 2001; 22: 50–52.
17. Tendero C, Tixier C, Tristant P, et al. Atmospheric pressure plasmas: a review. Spectrochim Acta B 2006; 61: 2–30.
18. Cheng SY, Yuen CWM, Kan CW, et al. Influence of atmospheric pressure plasma treatment on various fibrous materials: performance properties and surface adhesion analysis. Vacuum 2010; 84: 1466–1470.
19. Arai T, Freddi G, Innocenti R, et al. Acylation of silk and wool with acid anhydrides and preparation of water-repellent fibers. J Appl Polym Sci 2001; 82: 2832–2841.
20. Chen YU, Lin H and Ren Y. Study on Bombyx mori silk treated by oxygen plasma. J Zhejiang Univ Sci 2004; 5: 918–922.
21. Behnisch J, Hollander A and Zimmerman H. Surface modification of polyethylene by remote dc discharge plasma treatment. J Appl Polym Sci 1993; 49: 117–124.
22. Boenig HV. Plasma Science and Technology. New York: Cornell University Press, 1982.
23. Chiou MS and Li HY. Equilibrium and kinetic modeling of adsorption of reactive dye on cross-linked chitosan beads. J Hazard Mater 2002; 93: 233–248.
24. Ho YS and McKay G. Sorption of dye from aqueous solution by peat. Chem Eng J 1996; 70: 115–124.

11

Woven mulberry silk fabrics: Stretch and growth characteristics

Summary : There is a good relation between the stretch and growth properties and cover factor, thickness and warp crimp of the fabric. The stretch and growth properties of all the fabrics increase twofold in wet condition when compared to the dry condition. The maximum warp way stretch and growth properties of all the dupion silk fabrics increase twofold in wet condition compared to dry condition. The maximum warp way stretch and growth properties can be seen in both dry and wet conditions in the case of dupion silk fabrics as compared to that in all other fabrics. Inherent properties of dupion silk fibre, viz. tensile, crimp, yarn twist, and cover factor, may influence stretch and growth properties. Moreover, dupion silk fabrics show maximum growth property and require proper care while fabric processing and wash care. Silk fabrics are most desired for dry cleaning, considering above aspects.

11.1 Introduction

The deformations and applied forces which affect the mechanical properties of textile fibres are considered crucial and influence the fibre behavior during processing, and thereby determine the properties of a final product [1]. The water-soluble protein from the silkworm glands change into fibrous protein owing to either flow-controlled denaturation or mechanical denaturation and thus the fibre formation kinetics is relatively complicated. The behavior of silk yarn/fibre in case of mulberry against dynamic loading is having practical and theoretical importance. Silk is characterized fairly by its high strength and breaking extension, which combined to give much greater a work of rupture as compared to other fibres [2]. Silk fabrics woven by using a variety of raw silk such as mulberry exhibit different behavior when subjected to static and dynamic loading in wet and dry conditions [3]. The yarn properties and fabric geometry also cause different impact when subjected to external load under wet and dry conditions, influencing stretch and growth properties of a fabric. The availability of data or literature on stretch and growth properties of silk woven fabric is found to be scanty, as stated at ASTM standards [4]. The fabric stretch is of two types: comfort stretch and power stretch. The comfort stretch in wet and dry conditions, i.e. stretch and growth properties against fabric construction particulars, has been studied, considering that the formed

DOI: 10.1201/9781003735427-11

garment is expected to show minimum stretch and growth properties during regular use.

11.2 Technical details

Silk varieties used for the investigations are dupion silk, soft silk and taffeta silk at standard conditions [8].The fabrics so produced have been subjected to stretch and growth tests as per ASTM standards. Instron tensile tester of CRE type has been used. Stretch and growth properties have been studied under dry and wet conditions.

11.3 Stretch % of silk fabrics

The dupion fabric warp way stretch % in dry condition ranges between 3.22 and 4.28 and that of weft way ranges between 0.24 and 0.58. But the warp way and weft way stretch % ranges between 5.46 and 9.01 and 1.22 and 1.39, respectively, under wet condition, i.e. warp way stretch % in wet condition increased twofold than noticed in dry condition and in weft way it is 2.68 times more than that in dry state. In case of soft silk fabric, warp way stretch % varies from 2.04 to 3.42 and 4.16 to 7.22 in dry and wet conditions, respectively [8]. However, weft way stretch % shows variation from 1.03 to 1.77 and 1.57 to 2.45 in dry and wet conditions, respectively. Here also, it is found that warp way stretch % is increased twofold in wet condition as compared to that in dry condition and in weft way, it is found to be 1.42 times more. In case of taffeta silk fabrics, warp way stretch % varies from 1.71 to 2.98 and 4.01 to 6.85 in dry and wet conditions, respectively. But in weft way it ranges between 0.4 and 1.44 and 1.31 and 2.26 in dry and wet states, respectively. Hence, for taffeta silk also, warp way and weft way stretch % is increased by 2.47 and 3.27 times respectively in wet state as compared to dry state.

11.4 Growth % of silk fabrics

It has been observed that with dupion silk, growth % in wet state warp way increases 4.22 times to that seen in dry state, which can be due to highest warp crimp of the yarn. Also, weft way growth % is found to be 1.47 times more than that in dry condition, which has very less weft yarn crimp as compared to warp crimp. This increase may be attributed to highest warp crimp of the yarn. In case of soft silk fabrics, warp way growth % in wet condition is 2.95% more than in dry condition, while weft way growth %

is found to be 1.68 times more than that in dry condition. The taffeta silk fabrics show 3.48 times more warp growth % in wet condition as compared to that in dry condition. Similarly, weft way growth % in wet condition is found to be 1.90 times more than that in dry condition. It is concluded that all the fabrics have shown greater growth percentage than stretch percentage in wet condition, which may be attributed to more swelling and relaxation of fibre structure in wet state and also to hydrogen bond breaking inside chain links of the protein polymer structure of silk filament. The tensile, crimp and inherent properties of fibres used in fabric construction, such as fabric GSM, cover factor and yarn twist, have shown influence on stretch and growth properties of the fabric in dry and wet conditions [6]. Silk differs from other fibres by its considerable permanent setting in dry or wet states and at the same time it resolves to high immediate elastic recovery after mechanical conditioning [7].

11.5 Stretch and growth properties

Correlation technique has been used to statistically evaluate the effect of fabric construction particulars on stretch and growth properties of silk fabric. Under dry and wet states, the stretch and growth properties of silk fabrics can be analyzed as follows:

11.5.1 Dry state

The stretch in warp direction is negatively correlated to warp cover factor and positively correlated to weft cover factor. The growth along warp is negatively correlated to weft crimp factor. The higher cover factor of the fabric indicates higher compactness of the fabric and thus limits the stretch properties of the fabric. The crimp of the fabric depends upon density and diameter of the yarn; hence, increase in weft crimp leads to decrease in warp growth property of a fabric. Weft way stretch is negatively correlated to fabric GSM, weft linear density, weft cover factor and fabric thickness [8]. Similarly, it is positively correlated to picks/cm and weft twist. Higher fabric thickness and cover factor reduce space for movement of higher linear density of yarn in the fabric, thus reduces stretch property of a fabric. Weft way growth is negatively correlated to GSM, fabric cover factor and fabric thickness, and positively correlated to weft twist, i.e. the fabric cover factor and thickness have replicated effect on growth property in weft way of the fabric. Thus the cover factor controls the growth property of the fabric and increase in fabric density controls the thickness.

11.5.2 Wet state

The stretch in the warp direction in wet state bears a positive relationship to weft cover factor and weft linear density, and has negative relationship to weft cover factor and weft linear density, and negatively correlated to warp cover factor. This indicates that the increase in weft cover factor increases the fabric stretch property. In other way, increase in picks/cm will enhance stretch property of the fabric. Further, warp way growth is positively correlated to weft linear density and weft cover factor, and negatively correlated to weft twist, i.e. the warp way growth depends on weft denier and weft cover factor. In other words, warp way growth property can be increased by the increase in weft cover factor and picks/inch [8]. The weft way stretch is negatively correlated to weft linear density in wet condition, i.e. weft way stretch property of the fabric decreases by influence of higher weft yarn linear density. Weft way growth of the fabric is negatively correlated to fabric GSM, cover factor, and positively correlated to weft twist. On the other hand, increase in fabric cover factor and fabric thickness reduces the growth property of the fabric. Increase in ends/cm and picks/cm of the fabric increases the fabric cover factor and thickness.

References

1. Morton WE and Hearle JWS, Physical properties of textile fibers (The Textile Institute, Heinmann, London), 1975, 265.
2. Meredith R and Hearle JWS, Physical methods of investigating textiles (Interscience, New York, USA), 1959.
3. Juodsnukyte Daiva, Daukantiene Virginija Gutauskas, International journal of clothing science and technology, 20[1], 1989, 7–14.
4. Annual Book of ASTM standard (American society for testing and materials, West coshohocken, USA) 2001, 1199.
5. Tatsuki Matsuo, Indian journal of fibre and textile research, 31, 2006,142.
6. George Susich and Stanley Backer, Textile research journal, 21, 1951, 482.
7. Susich G and Zagieboylo W, Textile research journal, 23, 1953, 407.
8. Thimma Reddy G, Vijayakumar Kathari P, Aswatha Reddy, and Subrata Das, Indian journal of fibres and textile research, 2011, 248.

12

Enhancement of coloration and antimicrobial properties of silk fabrics through nanotechnology

Summary : The color strength of the fabric increases with the concentration of polyurethane acrylate + PEG 6000 binder. When polyurethane acrylate + PEG 6000 binder is used in ink preparation, it gives higher color strength than that used in printing paste. With the increase in concentration of titanium dioxide nano particles, the color strength and ultra violet protection increases, especially in the case of printing. Use of titanium dioxide in either printing paste or in ink preparation shows antibacterial properties good for Escherichia coli (G-) and Staphylococcus aureus (G+). The results obtained in case of printing are found to be better than those obtained in case of dyeing. There is improvement in overall fastness properties by using titanium dioxide nanoparticles in either ink preparation or in printing paste.

12.1 Introduction

Pigment printing is known to be the oldest and the easiest printing technique considering its simplicity of application [1, 2]. Owing to its merits such as versatility and ease of near final print at the printing stage itself, pigment printing enjoys wide popularity. During printing, it is necessary to consider formaldehyde emissions and clogging on the screens, which are demerits related to the binders used [3–5]. Initially, the study on the use of nanotechnology in textiles was undertaken by Nanotex, a subsidiary of the US-based Burlington industries [6]. Later more and more textile companies began to invest in the application of nanotechnology in textiles. Coating is a common technique used to apply nano particles onto textiles. The coating compositions that can modify the surface of textiles usually include nano particles, surfactant, ingredients and carrier medium [7]. Several methods have been explored to apply coating onto fabrics, including spraying, transfer printing, washing, rinsing and padding of these methods; padding is the most commonly used [8–10]. The nanoparticles are attached to the fabrics with the use of a padder adjusted at suitable pressure and speed, followed by drying and curing. The properties imparted to textiles using nanotechnology include water repellency, soil resistance, wrinkle resistance, antibacterial, antistatic, UV protection, flame retardation and improvement in dyeability. As there are various potential applications of nanotechnology in the textile industry, some

DOI: 10.1201/9781003735427-12

well-known properties imparted by nano treatment are critically highlighted in the paper. An aqueous UV curable binder of polyurethane acrylate has been synthesized based on polyethylene glycol 6000, with titanium dioxide nanoparticles either in ink preparation for pigment dyeing or in printing pastes for coloration of silk fabrics so as to improve their UV protection, antibacterial and wrinkle resistance properties.

12.2 Technical details

Low-weight silk fabric is treated with anionic detergent sodium carbonate at 45°C for 30 min., washed well and then air dried at room temperature. Polyurethane acrylate has been synthesized [11–13]. The silk fabric has been dyed with aqueous pigment inks. Flat-screen printing method has been used to apply the prepared printing paste onto the silk. Thermofixation has been done at 90°C for 3 minutes through the polymerization process. The color strength of the printed fabric has been evaluated [14]. Ultra violet protection factor values have been evaluated. Crease recovery angle has been measured. Antimicrobial activity against gram-positive and gram-negative bacteria has been evaluated. SEM investigations have been done on the printed and dyed silk fabrics. The rubbing, washing and perspiration fastness have been determined [26].

12.3 Evaluation of polyurethane acrylate

Most acrylated oligomers are based on polyether, polyester and expoxy resins, as the functionality of these condensation resins can be precisely controlled to 2 or 3 to prevent gelling and maintain low viscosity. At controlled reaction conditions and catalyst concentrations, diisocyanate could react with polyethylene glycol to form isocyanate-terminated polyurethane prepolymer. The polyethylene glycol reacts with the primary isocyanate group of isophorone diisocyanate leaving the secondary isocyanate group intact for subsequent reaction with hydroxy ethyl acrylate. From the infrared spectra, the formation of the polyurethane acrylate oligomer is evidenced by the emergence of strong absorption bands at 1244 cm^{-1} (C-O), 1459 cm^{-1} (CNC), 1533 cm^{-1} (N-CO), 1637 cm^{-1} (C=C), and 1721 cm^{-1} (C=O). It is also clear from the spectra that there is no absorption band at ~ 2274 cm^{-1} [15], which corresponds to NCO group. It shows that entire quantity of isophorone diisocyanate enters in the reaction but the end product is isocyanate free. All this confirms the occurrence of the addition reaction to get on the aqueous binder of polyurethane acrylate based on polyethylene glycol 6000.

12.4 Influence of polyurethane acrylate concentration on color strength value

Table 12.1 shows the influence of polyurethane acrylate binder concentration on the color strength of either screen-printed or pigment-dyed silk fabrics. It can be seen that as the concentration of PUA + PEG 6000 binder increases, the color strength increases. Moreover, greater color strength is obtained using PUA + PEG 6000 binder, than those obtained on using this binder in printing paste. For example, the color strength results are higher than those obtained on using this binder in printing paste. For example, the color strength values are 8.89 and 4.55 for silk fabric using 4% of PUA + PEG 6000 binder on the color strength of either screen-printed or pigment-dyed silk fabric is shown in Table 12.2. It is clear from Table 12.2 that as the concentration of titanium dioxide nanoparticles increases the color strength of the colored fabrics increases, especially in case of printing. The increase in color strength may be due to the increase in titanium dioxide nanoparticles which catalyzes the rate of polymerization reaction of the binder through the unsaturated bonds [26]. This leads to more fixation of the pigment and increases the color strength. Moreover, in case of using PUA + PEG 6000 binder in ink preparation, the color strength results are higher than those obtained in the case of using printing paste.

Table 12.1 Effect of PUA binder concentration on color strength of silk fabrics screen printed/pigment.

Dyed [26]				
Binder concentration (PUA + PEG 6000 binder)	Color strength			
	Pigment dyeing[1]			Printing[2]
	70°C	90°C	110°C	
0	1.85	2.1	2.3	1.58
2	5.5	5.53	6.1	3.87
4	8.64	8.89	9.25	4.55
6	6.6	7.92	8.82	4.43
6 (without wach)	6.83	8.75	8.95	4.56

[1] Time of fixation is 3 min.

[2] Fixation temperature 90°C and time of fixation 3 min.

Table 12.2 Effect of concentration of titanium dioxide nanoparticles on color strength of silk fabrics

Duration of fixation – 3 minutes; Temperature of fixation – 90°C [26]			
Binder concentration	Titanium dioxide concentration	Color strength	
%			ppm
		Dyeing	Printing
4	0	8.89	4.55
0	25	8.90	4.09
4	25	7.82	7.47
4	50	8.70	8.56
4	100	9.57	10.20

12.5 UV protection

As the inorganic UV blockers are nontoxic and chemically stable under exposure to both high temperatures and UV radiations, they are more preferable to organic UV blockers. Inorganic UV blockers are generally certain semiconductor oxides that include titanium dioxide, zinc oxide, silicon dioxide and aluminium oxide. Among these semiconductor oxides, titanium dioxide and zinc oxide are commonly used [16–20]. It has been found that nano-sized titanium dioxide and zinc oxide are more efficient for absorbing and scattering UV radiation than the conventional size particles and therefore are better to block UV [19]. Various studies on the UV protection treatment to fabric using nanotechnology have been conducted [16–18]. UV protection treatment for cotton fabrics was studied using sol-gel method [21]. A thin layer of titanium dioxide is applied on the surface of the treated cotton fabric which provides excellent UV protection; the effect can be detained even after 50 home launderings [21]. The effect of titanium dioxide nano particles with and without using 4% PUA binder on ultraviolet protection factor of either screen-printed or pigment or pigment-dyed silk fabric is shown in Table 12.3. It is clear from Table 12.3 that as the concentration of titanium dioxide nanoparticles increases the UPF of the fabrics increases, especially in case of printing silk fabrics [26]. The use of PUA + PEG 6000 binder alone also gives an UPF due to the presence of unsaturation bonds in this compound. Also the presence of titanium dioxide nanoparticles in either printing paste or in ink used for dyeing increases the UPF, and an excellent UPF rating is obtained due to the fact that nano particles have a larger surface area per unit mass and volume than the conventional materials, leading to increase in the effectiveness of blocking UV radiation. The light scattering in the case of small particles predominates at approximately one tenth of the wavelength of the scattered light. According to Raleigh's scattering theory, the scattering is strongly dependent upon the wavelength, and inversely proportional to the wavelength to the fourth power [16].

Table 12.3 Effect of titanium dioxide nano particle concentration on UPF of silk fabrics

Duration of fixation – 3 minutes; Temperature of fixation – 90°C [26]			
Binder concentration %	Titanium dioxide concentration ppm	Ultraviolet protection factor (UPF)	
		Dyeing	Printing
0	0	10	10
4	0	18.83	37.03
0	25	27.18	60.40
4	25	21.90	53.25
4	50	23.61	27.55
4	100	25.50	59.73

12.6 Antimicrobial property

Nanosized silver, titanium dioxide and zinc oxide have been used for imparting antibacterial properties [16, 17, 19, 22–25]. Metallic ions and metallic compounds exhibit sterilizing effect to a certain extent. A part of the oxygen in the air/water is supposed to be converted to active oxygen owing to the catalysis with the metallic ion, thus dissolving the organic substance to produce a sterilizing effect [19]. The effect of titanium dioxide nanoparticles concentration with and without using 4% PUA binder on antimicrobial properties of either screen-printed or pigment-dyed silk fabric is shown in Table 12.4. It is found that the use of PUA binder alone has no effect on the antimicrobial property, but upon using titanium dioxide in either printing paste or in ink preparation good results are observed for E.Coli (G−) and Staphylococcus aureus (G+). Also, the results obtained in case of printing are better than those in case of dyeing [26]. This may be due to the fact that titanium dioxide is a photo catalyst, once it is illuminated by light with energy higher than its band gaps the electron in titanium dioxide jump from the valence band to the conduction band, and the electron and electric whole pairs will be formed on the surface of the photo catalyst. The negative electrons and oxygen combine into the positive electric holes and water generates hydroxyl radicals. As both are unstable chemical substances, when the organic compound falls on the surface of the photocatalyst, it combines with oxygen and hydroxyl ions respectively and turns into carbon dioxide and water, which is known as 'oxidation–reduction' [18]. Common organic matters in the air, such as odorless molecules, bacteria and viruses get decomposed by the photo catalyst by means of the reaction.

Table 12.4 Effect of concentration of titanium dioxide nano particles on UPF of silk fabrics

Fixation – 3 minutes; Temperature of fixation – 90°C [26]					
Binder concentration	Titanium dioxide concentration	Inhibition zone diameter			
%	ppm	mm/1 cm sample			
		Dyeing		Printing	
		E.coli	S.aureus	E.coli	S.aureus
		(G^-)	(G^+)	(G^-)	(G^+)
4	0	0	0	0	0
0	25	19	23	27	30
4	25	15	16	16	18
4	50	16	17	17	19
4	100	18	20	20	22

12.7 Wrinkle resistance

The influence of titanium dioxide nanoparticles concentration with and without using 4% PUA binder on crease recovery of either screen-printed or pigment-dyed silk fabric has been studied [26]. The crease recovery improves with the increase in concentration of titanium dioxide.

12.8 Fibre surface

The SEM photographs of silk fabric, printed silk fabric using 4% of PUA + PEG 6000 binder, and printed silk fabric using PUA + PEG 6000 + 25ppm titanium dioxide nano particles are shown in Figure 12.1. The micrograph of the silk fibre shows a smooth surface as compared to that of the printed silk fabric using 4% PUA + PEG 6000 + 25ppm titanium dioxide, which appears to be covered by a layer of titanium dioxide. However, in case of 4% PUA

+ PEG 6000 binder only a thin coating is seen. This difference is due to the titanium dioxide nanoparticles, coated on the fibre surface of silk fabrics [26].

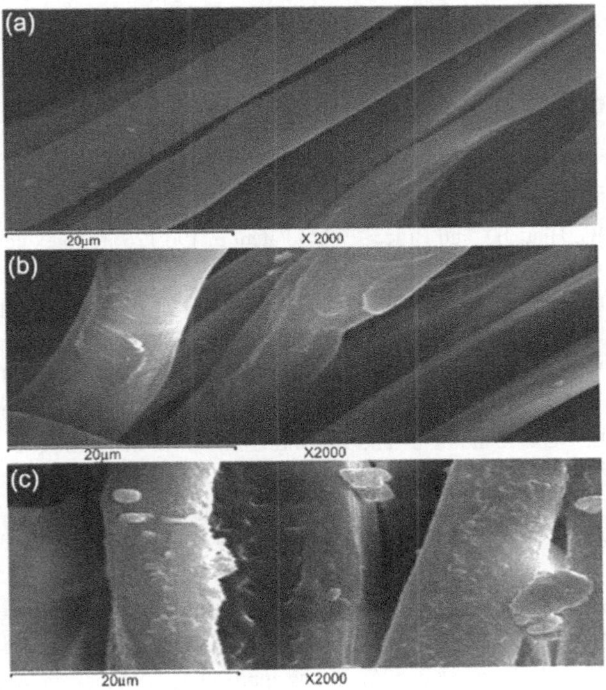

Figure 12.1 SEM photographs of (a) silk fabric (b) printed silk fabric using 4% of PUA+PEG 6000 binder, and (c) printed silk fabric using 4% of PUA + PEG 6000 binder + 25ppm titanium dioxide [26].

12.8 Fastness properties

The color strength and fastness properties of silk fabrics with PUA + PEG 6000 aqueous binder in either screen-printed or pigment-printed dyed silk has been studied. All the coloured fabrics have been characterized by the soft handle. It can be because of the presence of the long chain hydrocarbon present in the prepared binder PUA + PEG 6000. The colour strength value depends on both the nature of binder used and the concentration of titanium dioxide nanoparticle. It is clear from Table 12.6 that the overall fastness properties improve by using the titanium dioxide nanoparticles in either ink preparation or in printing paste [26]. This may be due to the fact that titanium dioxide enhances the unsaturation site of the vinyl group which is responsible for the fixation of dispersed pigment during thermofixation through the polymerization process. The fastness properties achieved through use of PUA + PEG 6000 aqueous binder are found to be good. But the colourfastness to washing, rubbing and perspiration ranges from good to very good by use of titanium dioxide nanoparticles.

References

1. Schwindt W and Melliand Textilber, 71(1990), 693.
2. Holme I. Review of progress in colour, 22(1992), 1.
3. Smith HG, Hiatt JD, and Bruce JE. US Patent 6, 196,126 (to Intex corporation, NC) 2001.
4. Shah KP, and Panchmatia PR. US Patent 5, 969,018 (to the BF Goodrich company, USA) 1999.
5. Hotton WT and Ronald WN. US Patent 5, 143,954 (to Rohm and Haas Company, USA).
6. Russel E. Nanotechnologies and the shrinking world of textiles, Textile horizons, 9/10(2002) 7–9.
7. Cramer Dean R, Ponomarenko Anatolyevna E, Laurent S, and Burckett JCTR. Method of applying nanoparticles, US Patent 6,645,569 (to Procter and Gamble company), 2003.
8. Small scale technology with the promise of big rewards, Technical textile international, 3(2003) 13–15.
9. Xin JH, Daoud WA, and Kong YY. A new approach to UV blocking for cotton fabrics, Textile research journal, 74(2004), 97–100.
10. Yeo SY, Lee JH, and Jeong SH. Preparation of nanocomposite fibres for permanent antibacterial effect, Journal of material science, 38(2003), 2143.
11. Yang J Wang Z, Zen ZG, and Chen Y. Journal of applied polymer science, 84(2002), 1818.
12. Aaoi T, Maenoto K, Kamiya A, and Misu H. US Patent, 4877711 (to Fuji Photo film co., ltd.) 1989.
13. El-Molla MM. Dyes and pigments, 74(2007), 371.
14. Lau KC. Dynamic response to color specifications, Journal of society of dyers and color, 111(5)(1995), 142.
15. Lee JJ, Chi KZ, Chang HH, and Ch YD. Journal of applied polymer science, 57(1995), 1005.
16. Burniston N, Bygott C, and Stratton J. Nanotechnology meets titanium dioxide, Surface coatings international, Part A, (2004), 179.
17. Sherman J, and Jonathan A. Nanoparticulate titanium dioxide coatings, and processes for the production and use thereof, US Patent 6653356 (to Fuess and Davidness), 2003.

18. Yang HY, Zhu SK, and Pan N. Studying the mechanisms of titanium dioxide as ultraviolet blocking additive for films and fabrics by an improved scheme, Journal of applied polymer science, 92(2003), 3201.

19. Saito M. Antibacterial, deodorizing, and UV absorbing materials obtained with zinc oxide (ZnO) coated fabrics, Journal of coated fabrics, 23(1993), 150.

20. Xiong MN, Gu GX, You B, and Wu LM. Preparation and characterization of poly(styrenebutyl acrylate) latex/nano-ZnO nanocomposites, Journal of applied polymer science, 90(2003) 1923.

21. Daoud WA and Xin JH. Low temperature sol-gel processed photocatalytic titania coating, Journal of sol-gel science and technology, 29(2004), 25.

22. Wong YWH, Yuen CWM, Leung MYS, Ku SKA, and Lam HLI. Selected applications of nanotechnology in textiles, AUTEX Research Journal, 6(1), 2006.

23. Lee HJ, Yeo SY and Jeong SH. Antibacterial effect of nanosized silver colloidal solution on textile fabrics, Journal of material science, 38(2003), 2199.

24. Yeo SY, and Jeong SH, Preparation and characterization of polypropylene silver nanocomposite fibres, Polymer international, 52(2003), 1053.

25. Daoud WA and Xin JH, Nucleation and growth of anatase crystallites on cotton fabrics at low temperatures, Journal of American Ceramic society, 87(2004), 953.

26. El Molla MM, El Khatib, El Gammal MS, Abdel Fattah. Nanotechnology to improve coloration and antimicrobial properties of silk fabrics, Indian journal of fibres and textile research, 36(2011), 266.

13

Influence of electric field on bivoltine mulberry and tasar fibres

Summary : The presence of critical field beyond which the electric modulus parameter decreases with the increase in magnitude of electric field, is shown by the changes in crystallite-shaped ellipsoid in two varieties of silk fibres in the presence of time-varying electric field. This is due to the increase in the strengthening of the crystallite regions in the polymer network. The influence of the crystallite region on the strength of the fibre in the presence of time-varying electric field can be considered quantitative data interpretation.

13.1 Introduction

Silk fibre comprises of crystalline and amorphous regions. There are cross-linked network along the length of fibre with the peptides. The crystal structure to be the beta-pleated sheets of the silk have been analyzed [1]. It is known that the external forces on the materials deform their size and shape. The X-ray and capacitance in a polymer film in the presence of electric field has been studied and found that there is no significant change in the layer spacing in polymer films [2]. Xu et al. [3] have reported the effect of electric field on block copolymer domain and found that there is significant lamellae orientation parallel to the electric field. Yaov Tsori [4] observed that electric fields on a polymer do change orientation and polymer network. Jong Hyoen et al. [5] have reported the existence of critical electrical field above the recombination of monomers. The present study is of importance since there is a general trend of electrospinning of silk from regenerated silk fibroin. This technique is most widely used technique to form artificial nanofibers and nanostructures [6, 7]. In the case of electrospinning techniques, the high voltages are applied to discharge the polymeric solution to form nanofibres. In fact, even a moderate voltage of the order of 100 V on the silk fibres creates micro stress in the silk fibres. It causes strain in the fibres, leading to change in Young's modulus of the fibres. This chapter highlights the aspect without going into the electrical conductivity issue in the case of two varieties of silk fibres, viz. the domestic mulberry – bivoltine (bombyx mori) silk and tassar (Antharaea pophia) silk fibres that are of commercial importance.

DOI: 10.1201/9781003735427-13

13.2 Silk samples

Bivoltine mulberry and wild tassar silk fibres have been used for investigation. The domestic and wild cocoon varieties have been collected from the germplasm and reeled following the standard procedure [8]. Firstly, the cocoons were cooked in boiling water at 100°C for 2 min to soften the sericin and then transferred to water bath at 65°C for 2 min. Lastly, the cocoons have been reeled in warm water (40–45°C) with a reeling equipment.

13.3 Recording of X-ray diffraction

A glass plate has been used to place the silk threads parallel to the length of the glass slides and then the two copper strips were placed at both the ends of the winding of the fibre. The distance between the copper strips is 0.02 mm and the arrangement used for *in situ* X-ray diffractometer recording is shown in Figure 13.1. The specific for recordings are 30 kV and 15 mA. Alternating voltages of 100, 150, 200 and 250 volts have been used with a step down transformer. The scattered X-rays have been scanned for these voltages and for two different silk fibre varieties [16]. Figure 13.2 depicts the X-ray recordings along with simulated whole powder patterns.

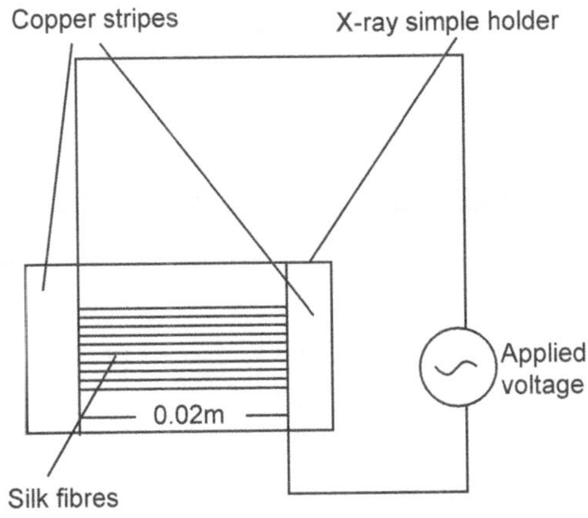

Figure 13.1 Experimental set up for the X-ray diffraction of silk fibres in presence of electric fields [16]

13.4 X-ray data analysis

W-H plot and whole powder pattern fitting analysis. For assessing the Interplanar spacing, crystallite size and lattice strain. Two/three prominent

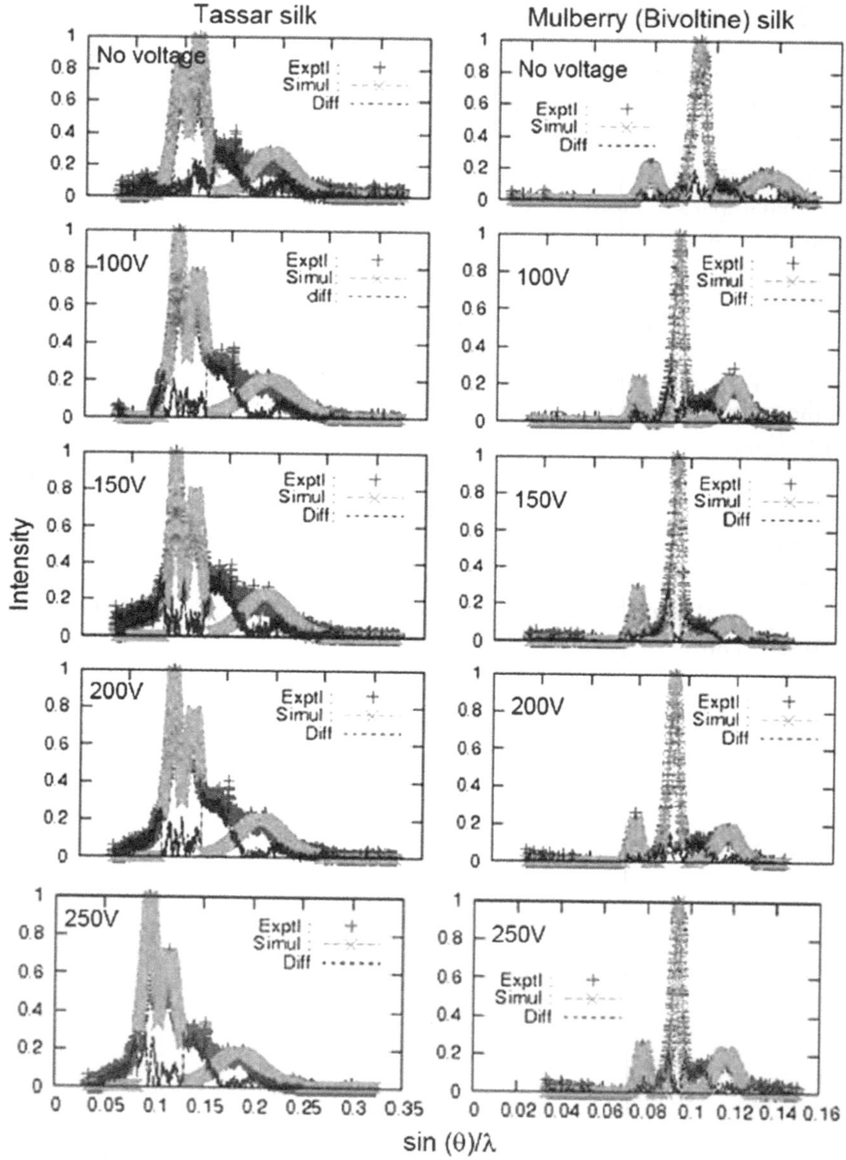

Figure 13.2 X-ray diffraction pattern for different magnitudes of electric fields in Mulberry (bivoltine) and tassar silk fibres along with simulated whole pattern fitting [16].

Bragg reflections have been seen in bivoltine and tassar silk fibres, respectively. Under the presence of electric fields, there is a slight change in the interplanar spacings of these reflections. The Williamson hall plot for computing the crystallite size and root mean square strain present in these fibres have been studied [16]. A whole powder pattern fitting based on single order line profile analysis has also been developed. An in-house program has been written for the computation of microstructural parameters [10–13]. These microstresses cause line broadening. In the presence of external electric field or pressure, it is shown that the lattice parameter can be expressed as a function of Bragg angle of corresponding plane (hkl), shown in Figure 13.3 below [14, 15].

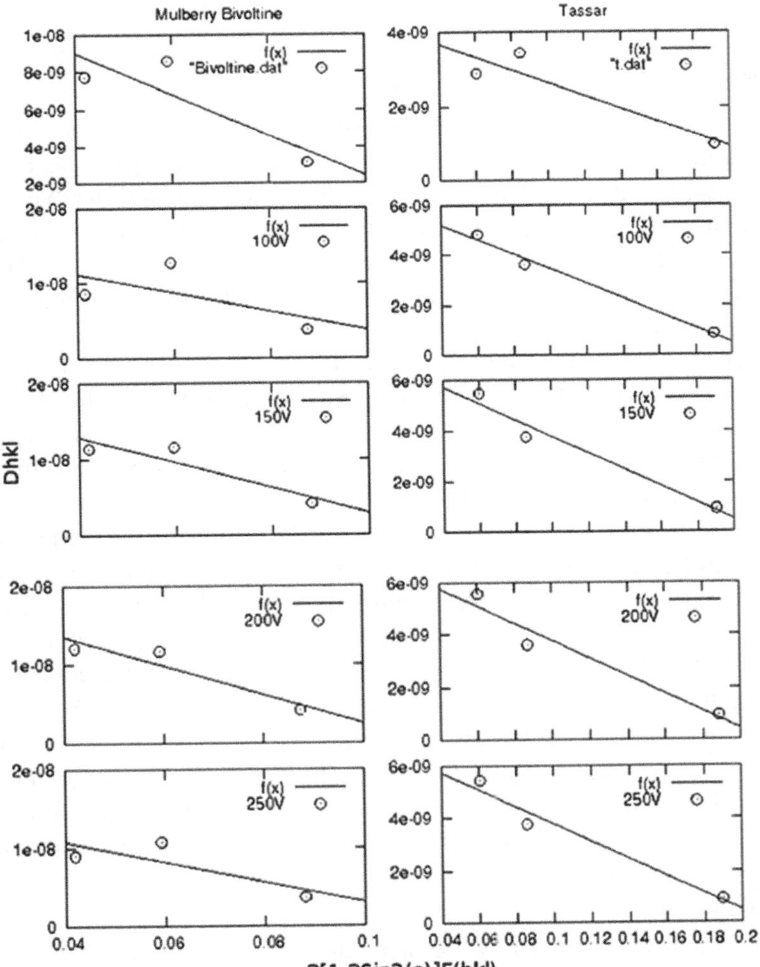

Figure 13.3 Variation in lattice parameter with [3(1 – 3sin$^2θ_{hkl}$)Γ(hkl)] for bivoltine and tassar silk fibres [16]

$$d(hkl) = M_o + M_l\,[3(1 - 3\sin^2\theta_{hkl})\Gamma(hkl)]$$

where M_o and M_l are the constants at a given electric field and are functions of microstress induced in the polymer. In fact these constants are a measure of elastic modulus of the polymer. A plot of the lattice parameter versus $[3(1 - 3\sin^2\theta_{hkl})\Gamma(hkl)]$ yields the parameters M_o and M_l. These are constants at a given electric field induced pressure and are functions of axial and radial stress components and hence elastic stiffness constants. These factors are computed as a function of electric field. The linearity of the above equation for bivoltine mulberry and tassar silk fibres for various fields is depicted in Figure 13.3.

13.5 Results of findings

Figure 13.2 clearly shows that in the presence of electric field, a significant change in X-ray diffraction pattern of silk fibres is seen. Changes are seen in the position of Bragg's reflections and thus results in considerable change in the unit cell dimension. As the field is applied along the fibre length, a drastic change occurs in the direction at right angle to the length of the fibre as indicated by the shift in the position of Bragg reflections. This linear strain coefficient varies from 0.2% to 1.1% due to change in applied electric field. Further an estimation of the crystallite size and root mean square strain in these fibres shows that there is a change in these parameters, leading to a more order of polymer network [16]. In fact, there is an evidence of existence of critical field beyond which these parameters more or less decrease. Williamson hall method is more meaningful for a relative comparison. For an absolute estimation of crystallite size, the whole powder pattern fitting technique has been used for all these samples. The details of computation have been reported in earlier papers [10–13]. In fact the results presented are the average of the crystallite size and lattice strain obtained for different asymmetric column length distribution functions, as mentioned earlier [11]. It is evident that there are significant changes in these parameters with electric field. For a better perspective, these values have been projected in a two-dimensional plot which results in the crystallite shape ellipsoids. The variation shape ellipsoids with electric field for both silk varieties are shown in Figure 13.4. These changes in the crystallite shape ellipsoid computed using broadened reflections are elastic in nature as supported by the fact that they vanish once the electric field is switched off.

This is also supported by the fact that there is a shift in the position of Bragg reflections and hence the strains which are in the range of 0.2–1.1%. These extrinsic strains are also elastic in nature. For the range of elastic strains used in the experiments, the change is reasonable which can be accounted for by the expected rearrangement of polymer network in the fibres. The maximum percentage of change in the crystallite area due to the presence of electric field in bivoltine and tassar fibres is 3775 A^2 and 1240 A^2, respectively. The changes in the values of M_0 and M_1, which are a relative measure of stiffness constants of the fibres, are due to the fact that there is residual porosity in the samples. The presence of electric field gradient results in varying strain gradient in the sample. The width of electric field gradient contributes to the parameters M_0 and M_1. The axially symmetric electric field has not been considered for calculations. The observed changes in the shape ellipsoid arise due to deformation of samples from the varying electric field which is equivalent to a non-hydrodynamic compression. There are two factors contributing to the increase modulus parameters: the presence of varying electric field and the polymer network (crystallite) strengthening.

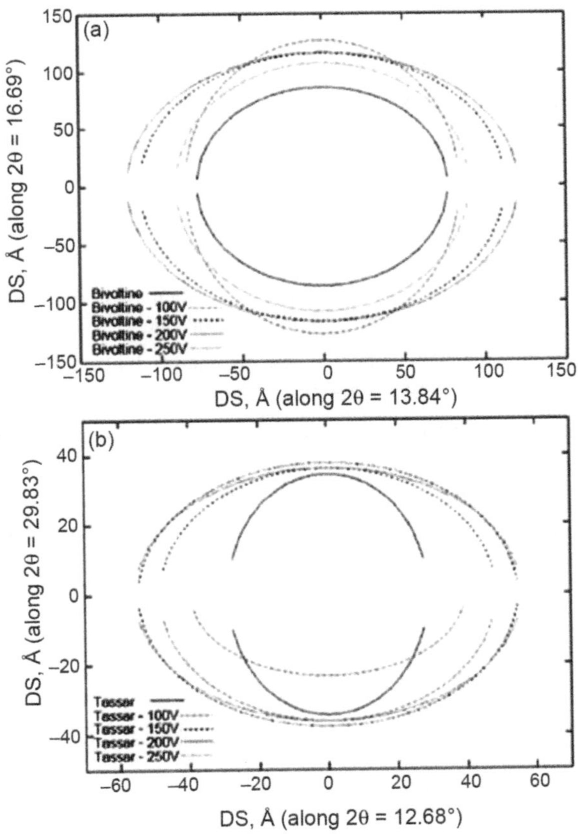

Figure 13.4 Variation in crystallite shape ellipsoids in the case of (a) Mulberry (bivoltine) silk fibre and (b) Tassar silk fibre in different electric fields [16]

References

1. Marsh RE, Corey RB, and Pauling L. Biophysics Acta, 16(1995), 1.
2. Poulsen M, Sorokin AV, Adenwalla S, Ducharme S, and Fridkin VM. Journal of Applied Physics, 103(2008), 034116.
3. Xu T, DeRouchey J, Thum-Albrechet T, Russel TP, and Kolbv R. Macromolecules, 37(7)(2004), 2538.
4. Yoav Tsori. Reviews in Modern Physics, 81(2009), 1471.
5. Jong-Hyeon J, Terashima D, Kimura C, and Aoki H. Japanese Journal of Applied Physics, 50(2011), 04DK09.
6. Jian-Feng Zhang, Dong Zhi Yang, and Jun Nie. Polymer International, 57(2008), 1194.
7. Audrey Frenot, Ioannis S, Chronakis. Current Opinion Colloid Interface Science, 8(2003), 64.
8. Annadurai V, Subramanyam, Gopalkrishnan R, and Somasekar. Journal of Applied Polymer Science, 79(2001), 1979.
9. Williamson GK and Hall WH, Acta Metallargica, 1(1953), 22.
10. Somashekar R, Hall IH, and Carr PD. Journal of Applied Crystals, 22(1989), 363.
11. Hall IH and Somashekar R. Journal of Applied Crystals, 24(1991), 1051.
12. Divakara S, Madhu S and Somashekar H. Pramana – Journal of Physics, 73(5)(2009), 927.
13. Souza FG (Jr), Soares BG, Mantovani GL, Manjunath A, Somashekarappa H, Somashekar R, and Siddaramiah. Polymer, 47(6)(2006), 2163.
14. Singh AK, Liermann HP, Saxena SK, Mao HK, and Usha Devi S. Journal of Physical Condensed Matter, 18(2006), S969.
15. Singh AK, Liermannb HP, and Saxena SK. Solid State Communication, 132(2004), 795.
16. Siddaraju GN, Ananda HT, Somashekarappa H, and Somashekar R. Effect of electric field on mulberry(bivoltine) and tassar fibres, Indian Journal of Fibres and Textile Research, 37(2012), 347.

14

Torsional rigidity of mulberry and non-mulberry silks

Summary : Bivoltine silk is considerably more brittle than multibivoltine silk. The denier of raw silk also has a significant influence on the brittleness of the raw silk. In bivoltine and multivoltine silks, the coarser deniers show greater brittleness. For any specified denier, the brittleness of non-mulberry silk is seen to be significantly different from that of mulberry. The brittleness of Muga is significantly lesser than tasar or mulberry silk. The brittleness of raw mulberry silk is significantly greater than either tasar or muga. The main factors contributing to the brittleness of any yarn are the filament denier and the shape of cross section. For a given denier, i.e. (40/42), the number of filaments in the cross section of bivoltine silk is because of the difference in the filament denier. Therefore, the area of cross section is more in the case of multivoltine silk, due to which the yarn is able to take up more twist, resulting in less brittleness compared to bivoltine silk, though the filament cross section is same in both the cases. The same applies to tasar and muga silk with respect to the number of filaments in the cross section. Considering the shape of the filament cross section, it is found to be triangular with rounded edges in the case of mulberry silk, while in the non-mulberry silk, muga silk has a triangular cross section with blunt edges. Tasar has a triangular elongated cross section with sharp edges. Therefore, though the number of filaments in the cross section of non-mulberry silk is less than that of mulberry silk, the shape of the filament cross section is perhaps contributing to the reduced value of brittleness. But, an elaborate investigation of the filament structure can provide more information on this aspect. Hence, non-mulberry silk is able to take more twists without being affected. So too, multivoltine raw silk has a better ability than bivoltine raw silk in taking the twist without affecting its strength.

14.1 Introduction

During their manufacture and use, textile materials are subjected to different kinds of deformations, such as tensile, shear and bending compression, which influence their serviceability. An understanding of the mechanical behavior of fibres subject to various processing treatments enables prediction of their ability to maintain their integrity during processing. Such information is useful to design equipments as per the limitations of materials to be handled [1]. Despite fibres being subjected to complex deformations during use and processing, many studies have focused on tensile properties. Knowledge of the shear behavior and fatigue life of fibres is also necessary since the fibres are not subjected to tensile stress alone during processing or end use.

DOI: 10.1201/9781003735427-14

Twisting is one of the important processes in yarn manufacture and during this process the yarn undergoes shear deformation. The number of twists or turns required to rupture a fibre (breaking twist) represents its maximum resistance to shear, when the tensile stress necessary to prevent the formation of loops and kinks is kept to a very minimum value [2]. The most convenient measure of this maximum resistance to twist is given by breaking twist angle (BTA) which is the compliment of the twist angle. The BTA can be characterized by the parameter (a) in units of radians or degrees. It is used as a measure of brittleness by many researchers under various studies [1, 2, 4–7]. A large BTA value corresponds to significant brittleness and in that low extension has been reported to be accompanied by a high BTA. BTA is the angle through which the outer layers are sheared and this could be used to measure a characteristic property of the fibre material. The chapter highlights the ability of silk yarns to take up twist without deterioration using the breaking twist angle for the first time. Comparisons have been made on the BTA in air of multivoltine and bivoltine raw silk and study of the interrelationship between denier and BTA of multivoltine and bivoltine raw silk, and also BTA in air of mulberry and non-mulberry raw silk varieties, namely, multivoltine, bivoltine, muga and tasar.

14.2 Technical aspects

The tenacity of raw silk increases from 3.6 g/den to 4 g/den in the case of multivoltine silk and from 3.6 g/den to 3.9 g/den in the case of bivoltine silk with the increase in the denier of raw silk from 16/18 to 40/42 [9]. The tenacity of tasar raw silk is 2.15 g/den whereas the tenacity of muga raw silk is 3.19 g/den for the same denier, i.e. 40/42. Studies carried out earlier have addressed the BTA of a monofilament or single fibre, while the present study is on raw silk comprising a bundle of single filaments, owing to its practical relevance. BTA in degrees was calculated using the following relationship:

$$a = \tan^{-1} (L/\pi dn) \qquad \text{...(14.1)}$$

where a is the BTA in degree; L is the length of sample in cm; π, the constant; d is the diameter of the sample in cm; and n is the number of turns at break.

The diameter of the sample can be calculated using the following formula:

$$d = \frac{4\mu \times 10^{-5}}{\pi\rho} \text{cm}$$

where d is the diameter in cm; μ, the linear density in tex; ρ is the mass density of fibre/yarn; and π is the constant. L is kept constant. Therefore, Eq. (14.1) gets reduced to

$$a = \tan^{-1}(1/dn)$$

The diameter (d) of raw silk of a given linear density is constant for any given type of silk. Thus a is equal to $\tan^{-1}(1/n)$.

This implies that the BTA (a) is inversely proportional to the number of turns at break 'n' and the number of turns at break can thus be used as an indicator of BTA. It is thus appropriate to use 'n' as an indicator of the brittleness of the yarn. Higher the number of turns at break, the lower is the BTA and thus the brittleness.

Results of findings

The data pertaining to number of turns at break of mulberry and non-mulberry raw silk have been obtained. For multivoltine raw silk, the average number of turns at break raw silk reduces from 283.6 to 180.45 and from 265.85 to 180.85 for bivoltine raw silk with the increase in the denier of raw silk from 16/18 to 40/42. For raw tasar silk, the average number of turns at break is 221.60 while in the case of muga silk it is 236.75 for the same denier, i.e. 40/42.

The data on number of turns at break were statistically analysed [9]. The data on the number of turns at break of multivoltine and bivoltine raw silk of all the deniers were subjected to two way analysis of variance. It is observed that for the deniers under study the number of turns at break of bivoltine raw silk is at 1% level of significance.

To study the interrelationship between denier and number of turns at break of multivoltine and bivoltine raw silk, the correlation and regression analysis were conducted. The correlation coefficient (r) in the case of multivoltine and bivoltine is (−0.948) and (−0.877), respectively. The regression equation between denier and number of turns at break for multivoltine is $Y = 350.253 - 4.092X$ and for bivoltine it is $Y = 315.219 - 3.354X$, where X is the denier and Y is the number of turns at break. The results show a significant (negative) correlation between raw silk denier and number of turns at break in the case of both multivoltine and bivoltine raw silk. The best fit line plot for multivoltine and bivoltine raw silk is given in Figures 14.1 and 14.2.

The results of one way analysis of variance of number of turns at break of mulberry and non-mulberry raw silk (multivoltine, bivoltine, muga and tasar) have been obtained [9]. From the results of one way analysis of variance and on comparison of difference between means with the critical difference, it is observed that the number of turns at break of mulberry and non-mulberry varieties of raw silk for a given denier, is significantly different for 5% level of significance. Thus the following conclusions could be made:

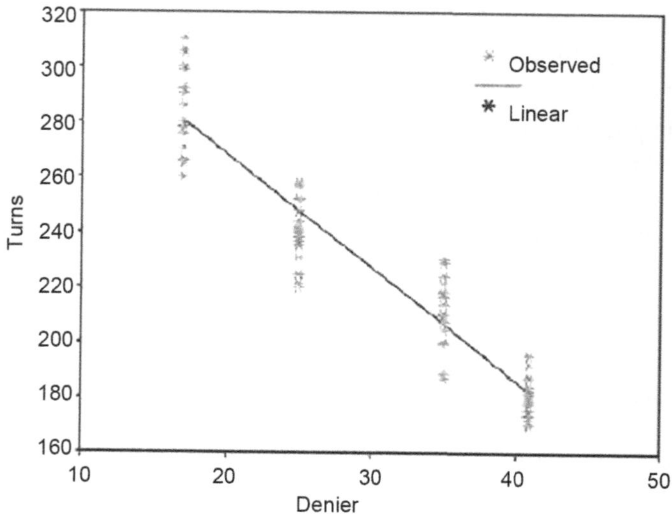

Figure 14.1 Best fit plot for multivoltine raw silk [9]

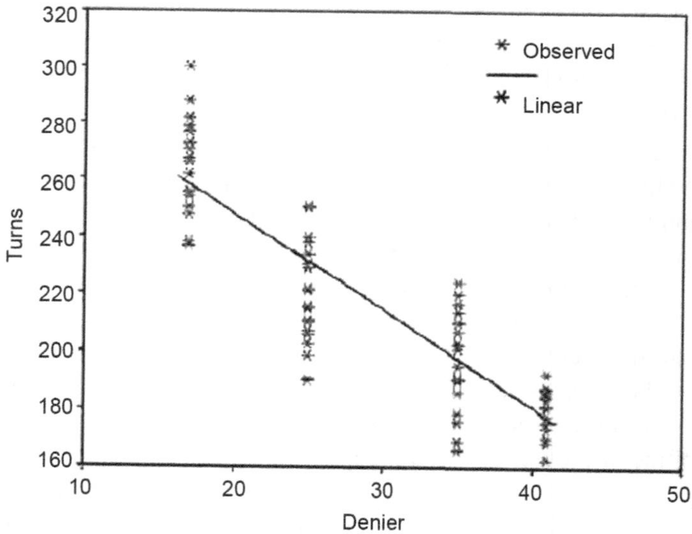

Figure 14.2 Best fit plot for bivoltine raw silk [9]

(a) For multivoltine, the number of turns at break is significantly higher than that of bivoltine, indicating that the former is less brittle than the latter.

(b) For raw non-mulberry silk, the number of turns at break is significantly greater than that of raw mulberry silk, indicating that raw non-mulberry raw is less brittle than raw mulberry silk.

(c) Muga is significantly less brittle than tasar among the non-mulberry varieties.

(d) As the yarn coarseness increases, its brittleness also increases.

(e) The rating of brittleness of various raw silk varieties in the descending order places bivoltine on the top followed by multivoltine and then tasar and muga at the bottom.

References

1. Ellison MS, Zeromain SH, Alger KW, Aboul Fadl, and Soler TM. Polymer engineering science, 29(24)(1989), 1738.
2. Zeronian SH, Xie Q, Buschle-Diller G, Holmes SA, and Inglesby MK. Journal of Textile Institute, 85(3)(1994), 293.
3. Ellison MS, and Lendgren HP, Textile Research Journal, 48(12)(1978), 692.
4. Aboul Fadl SM, Zeronian SH, Kamal MM, Kim MS, and Ellison MS, Textile Research Journal, 55(8)(1985), 461.
5. Zeromain SH, Buschle-Diller G, Fisher LD, Alger KW, Holmes SA, Slaven PA and Bertoniere NR, Textile Research Journal, 63(8)(1993), 488.
6. Selvakumar. A study on the effects of the chemical and physical treatments on the properties of silk yarn, PhD Thesis, Anna University, 1995.
7. Booth JE. Principles of textile testing, 3rd edition (Butterworth, London), 1982.
8. Jameela Khatoon, Design and fabrication of gadget to determine the breaking twist angle in air of textile fibres/ yarns on universal tensile testing machines such as instron/Qtest I, Project report (Bangalore university)1999.
9. Jameela Khatoon, Vasumathi BV, and Arindam Basu, Studies on torsional rigidity of mulberry and non mulberry silks, Indian Journal of Fibres and Textile Research, 37(2012), 353.

15

A qualitative and quantitative comparison of sericin from various sources

Summary : The quality of sericin differs with the source as well as the method of extraction. The best quality and quantity of sericin is achieved from woven silk fabric, whereas the poorest quality is obtained from silk waste. As the extraction is faster and greater in IR heated machine, it is much more efficient compared with HTHP machine. The drying technique influences the surface morphology. Spray drying yields globular aggregated particles while freeze drying gives planar sheet like particles. Molecular weight of sericin could vary from 20 kDa to 200 kDa depending on the source or method of extraction. Higher molecular weight sericin is soluble in only hot water while low molecular weight sample dissolves easily in water. Spectroscopic and other analysis shows that the sericin does undergo some degree of damage extraction process. It is necessary to determine the influence of this degradation on the performance properties. The results can be utilized to evolve a method for evaluating the quality of sericin protein obtained from different origins or methods of extraction and processing.

15.1 Introduction

Silk comprises mainly of fibroin and sericin, along with very small amounts of waxy substances, mineral salts and coloring matter [1]. The fibroin and sericin constitutes about 75% waxy substances, while mineral salts and coloring matter account for 25% of total silk weight, respectively [2]. Due to its primarily amorphous nature, greater water solubility and role of gum binder, sericin helps to maintain the structural integrity of the cocoon [3]. Sericin is predominantly composed of serine (about 32%), followed by aspartic acid (16.8%) and glycine (8.8%), and hence has a high concentration of hydroxyl groups [4]. Sericin has long been considered as a waste by product and generally unutilised. But there have been continual efforts to recover and reuse it as a natural biopolymer in various applications. Recently, it has been found that sericin shows several important useful properties such as antioxidant, antityrisonase, UV-absorbing and moisture-absorbing properties [5–7]. Sericin can be used as a finishing agent for natural or manmade textiles. The cosmetics industry is using sericin in skin care products. It can also be a valuable natural ingredient in food industry. Fibroin and sericin have been used for biomedical polymeric applications [8, 9]. Several processes like extraction with water at high temperature and high pressure (HTHP),

DOI: 10.1201/9781003735427-15

boiling off in soap, alkalis, acidic solutions, aliphatic amines, enzymes, and biosurfactants have been used for degumming of silk [10–15]. Recovery of sericin from soap alkali bath has been attempted using micro, ultra, and nanofiltration methods [16]. Drying has been carried out using spray drying, freeze drying, or tray drying [17–19]. Other methods of purification include precipitation of ethanol for lab-scale applications [20]. The properties of sericin obtained depend upon the method used for its extraction and recovery [21]. During degumming process sericin is fully or partially hydrolyzed, and is solubilized in degumming medium. Therefore, there is a need to develop extraction processes for sericin that are energy efficient, produce no chemical discharge and cause minimum damage to sericin. Recently, different forms of energies have become available for laboratory applications in textile industry. Infrared dyeing and drying machine works on the principle of radiation heating where heat is transferred in the form of electromagnetic waves directly to the centre of material without heating the surrounding air [21]. It is reported that IR heating has higher thermal efficiency and faster heating rate as compared to convection heating, so it is gaining popularity. A comprehensive analysis of the extraction conditions is important because these conditions can affect the final properties of sericin. In case of sericin, the extraction conditions should be carefully controlled since it is highly susceptible to heat. Extraction parameters such as temperature and time of treatment are known to affect the extraction efficiency and quality of sericin. The yield is crucial as it determines the economic value of the sericin extraction technique. Sericin has been extracted from four different sources of mulberry silk and the influence of different factors on extraction of sericin using infrared dyeing machine and HTHP machine have been investigated. The sericin extracted has been characterized and compared with the standard one.

15.2 Technical details

For sericin extraction, the silk obtained from various sources like silk cocoon, flat silk, reeling silk waste and woven silk fabric have been used. The quantity and purity of sericin from these sources vary since they represent various stages of processing of silk [37]. The purest form of silk is the silk cocoon. But, cocoons are rare and very costly. A special technique is used to produce silk flats where mature silk worms are kept on a spinning bed in place of a cocooning frame. The inclined spinning bed is rotated every few minutes to prevent the worm from spinning till it starts extruding silk on a flat surface thus flat sheets of silk are produced. They also yield a relatively pure form of silk. Silk waste is the silk left over after most silk has been reeled from the cocoon. The part that cannot be reeled is known as the silk waste. It contains insect parts, is difficult to clean and is preprocessed, so quality of silk is poor. Silk waste is a byproduct of the silk processing industry, relatively cheap and abundantly available. All silk fabrics are subjected to a degumming process for removing sericin from it. Therefore, this is the most obvious and abundant source for collection of sericin. Analytical grade chemicals, sodium carbonate, sodium hydroxide, copper sulphate, sodium potassium tartararte, bovine serine albumin, and Folin Ciocalteau's phenol reagent have been used for protein estimation. Deionised water has been used for all extractions. The following steps have been involved in the extraction and characterization of sericin:

(a) Extraction of sericin
(b) Determination of quality and quantity of sericin
(c) Preparation and characterization of sericin powder
(d) X-ray diffraction
(e) Spectroscopic analysis
 - Fourier transform infrared spectroscopy
 - Fluorescence spectroscopy
 - Circular dichroism spectroscopy
(g) Thermogravimetric and SEM study
(h) SDS-PAGE analysis

The various sources from which sericin have been extracted are cocoons, flats, silk waste and woven fabric. Qualitative and quantitative evaluations have been done on the extracted liquor by means of spectroscopy and protein content determination, respectively. The results have been benchmarked against standard sericin obtained from sigma.

15.3 UV spectroscopy

Figure 15.1 depicts the UV spectra of standard sericin and sericin extracted from different sources. At a bandwidth of 275.40 nm, the characteristic peak, caused due to the absorption of amino acids, is noticed in the standard sericin sample (A) and in the test samples (B-E). In the ultraviolet region, the proteins are strongly absorbed due to primarily the peptide bonds and aromatic acids. The aromatic amino acids like tryptophan, tyrosine, and phenylalanine are known to absorb UV radiation in the range of 260–290 nm due to $\geq$ * transition; and these absorptions can be used to assess the quality of sericin protein. Although all samples of sericin show the characteristic protein peak at 275.4 nm, the shape of the absorption spectrum is found to be different for different samples. [37]

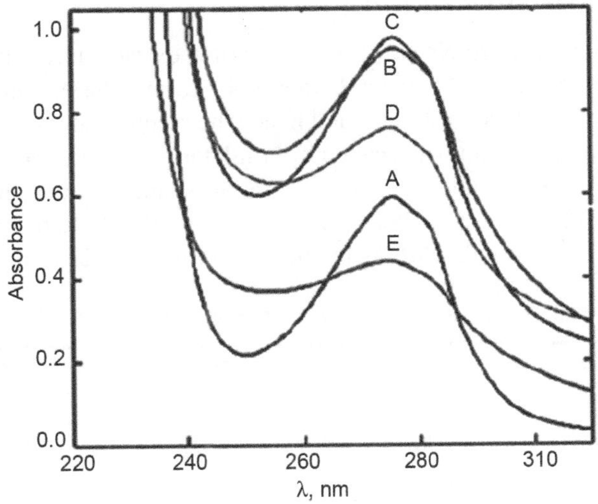

Figure 15.1 UV spectra of sericin extracted from various sources (100°C, 40 min) [37]

A – standard sericin ; B – Cocoons; C – Fabric; D – Flats; E – Waste

Sericin from the cocoons (B), silk flats (D), and silk waste (E) exhibit a broadening of the peak, as compared to the standard sericin, indicating some changes in the properties of protein. Sample C extracted from fabric shows a spectrum closest to that of the standard. This trend is endorsed by the values of yield (%) and A-ratio (A_{280}/A_{260}) for various samples given in Table 15.1. Maximum yield of 28% is observed with silk fabric. Slightly less yield is obtained with cocoons and silk flats. Minimum yield of 22% is obtained with silk waste. It is observed that A-ratio of sericin extracted from different sources at same extraction conditions is different, indicating that the source has a great impact on the quality of sericin. The order of source based on A-ratio is as follows:

Standard sericin > fabric > cocoons > flats > waste

Standard sericin shows the highest A-ratio of 1.73, which is close to the ideal value of 1.8. The sericin quality is better at higher the A-ratio. The next highest value of 1.35 is obtained for sericin extracted from silk fabric. The value of sericin extracted from cocoons also did not match the ideal one. The least A-ratio of 1.09 is exhibited by sericin extracted from silk waste, which shows certain changes or damage in sericin during the process of silk reeling. It gives a broad peak in UV and a corresponding low value of A-ratio. Hence, silk fabric gives the highest yield and quality of protein amongst all sources.

15.4 Protein content

Table 15.1 shows the protein content of the sericin extracted from the different sources. Standard sericin is available in powder form, whereas other samples are available in the form of liquor, and hence the results cannot be correlated. The protein content in liquor extracted from silk fabric is found to be maximum (31.62 mg/mL), while liquor extracted from fabric and cocoons show protein content of 29.12 mg/mL and 28.2 mg/mL, respectively. Protein content in the liquor extracted from silk waste is found to be minimum (23.2 mg/mL).

Table 15.1 A – ratio and protein content of various sericin samples [37]

Source of sericin	A – ratio	Protein content mg/mL	Yield %
Standard	1.73	–	–
Silk fabric	1.35	31.62	28
Cocoons	1.25	28.20	25
Silk flats	1.12	29.12	26
Silk waste	1.09	23.10	22

The silk fabric is considered to be the best source considering both the quality and quantity of sericin extracted from different source. Being abundantly available, it provides more scope for future investigations.

15.5 Sericin extraction with various energy sources

The conventional method of silk degumming involves the use of soap and alkali. The alkali results in degradation of sericin protein, but the industry continues to use this technique since it is economical and sericin recovery is not consideration for most processers. This technique is very polluting as it results in addition of large amounts of soap and alkali into the effluent stream. More critically, it is extremely difficult to recover sericin from these effluent streams. In order to maximize the recovery of sericin and minimize the damage to protein, extraction by water is the best option. Conventionally, HTHP machines are available in process houses and have been used by some researchers to separate sericin from the fabric [26]. However, some recent studies on extraction of sericin from silk waste show that HTHP extraction is not efficient and damages the protein [27]. In the current study, infra red heating is used for the extraction of sericin from silk fabric and results are compared with HTHP method. Figure 15.2 shows the yield of sericin obtained from the extraction of fabric for different time and temperature of extraction, using IR and HTHP machine. Yield of sericin at 15 min of extraction time in

IR method at 100°C is found to be 12.6%. Yield is found to increase slightly with increase in time with maximum reaching up to 14.6% at 150 min. Further increase in temperature to 120°C does not result in increase in yield.

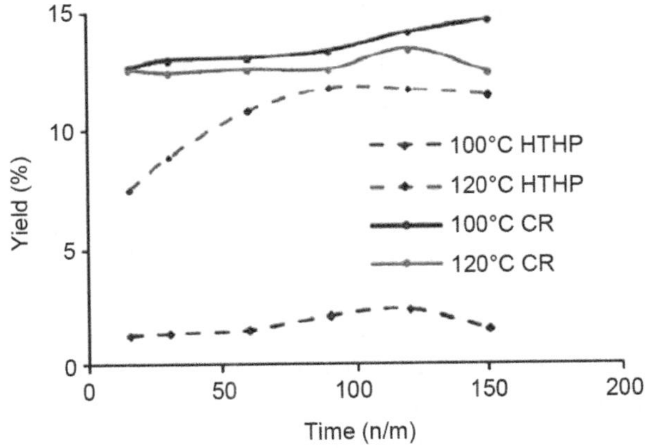

Figure 15.2 Influence of temperature and time [37]

Trend observed in HTHP extraction is, however, quite different. Yield in all cases is much lower than that obtained in IR method. Effect of temperature is much more prominent in HTHP as compared to that in IR heating bath. Yield at 100°C is very low (3%) even after 120 min of extraction. Yield is found to increase with time as well as temperature of extraction. Maximum yield of 12% is obtained at 120°C for 90 min of extraction. These observations clearly show that IR heating is a much more efficient method of sericin extraction as compared to HTHP heating. The yield is higher and time and temperature required are also less. This finding is similar to results reported earlier [23]. The difference between yield from IR and HTHP method is likely due to the principle of heating employed in the two cases. It can also be seen that IR is much more efficient than HTHP in removing sericin from fibroin since the bulk of sericin is removed within 15 minutes. Since sericin, like all proteins, is prone to degradation at high temperatures, 15 min at 100°C is taken as the suitable procedure for the extraction of sericin from silk fabric. The UV spectra of standard sericin and of that extracted by HTHP and IR machine show the characteristic peak at 275.4 nm. However, slight peak broadening is observed for samples B and C, thus suggesting some changes in the sample. Also, higher A ratio (1.35) which is indicative of protein quality is observed for the sericin sample extracted in the IR machine as compared to that extracted in HTHP (Table 15.2). The results suggest that the source of energy, i.e. method of heating, has a very important role in the quantity of protein obtained. As a lower temperature is desirable to prevent denaturation of proteins, the findings indicate that extraction in IR machines at 100°C for 15 min is the best method of extraction.

15.6 Sericin characterization

Better quantity and quality of sericin is obtained from fabric at 100°C for 15 min using infrared heating. Therefore, this liquor is converted to powder, and the characterization data for its composition, spectroscopic and thermal properties and molecular weight are generated. Results are benchmarked against the standard sericin sample and presented in the following paragraphs. The sericin powder extracted from fabric and standard sericin sample are characterized in terms of moisture, ash, nitrogen and protein content and the results are compiled in Table 15.3. Moisture content of the standard sericin sample is found to be slightly higher than the sericin extracted from the fabric. The ash content of standard sample is also found to be higher than the sericin sample recovered from fabric. This may be because a different method of extraction may have been used for the standard, resulting in higher residual matter. Gulrajani et al. have reported that ash content can vary from 0.8% to 5.2% in sericin extracted from cocoons using HTHP and alkali method, respectively [18]. Wu et al. report an ash content of 4.2% in sericin extracted from cocoons using HTHP method. The nitrogen content of the sericin powder is also estimated and the protein content is calculated by multiplying the value by 6.25. Sericin sample obtained from the fabric shows higher nitrogen content β (14.13%) and hence higher protein content (88.31%) as compared to the standard sericin sample. Nitrogen having identical values (14.65%) has been reported earlier [28].

15.7 X-ray diffraction analysis

The crystalline structure of sericin samples has been investigated using X-ray diffraction. Three types of conformations have been proposed for silk protein [28, 29]. The glandular state prior to crystallization is called silk I. Silk II is the spun silk state which consists of β sheet secondary structure and silk III (an air/water assembled interfacial silk) is a helical structure. The main diffraction peaks of silk I are present at around 2□ = 12.2° and 28.2°, while silk II are present at about 2 □ = 18.9° and 20.7°. The XRD curves of sigma sericin and sericin extracted from fabric exhibit a strong diffraction peak around 2□ = 20.5°. It is observed that the peaks are broad, indicating that the powders are amorphous. XR diffractograms are found to be similar for all sericin samples including the reference sample. The earlier investigations have reported similar results [28].

15.8 FTIR analysis

The proteins exhibit characteristic vibration bands in the range of 1630–1650 cm^{-1} for amide I (C-O stretching), 1540–1520 cm^{-1} for amide II (secondary N-H bending) and 1270–1230 cm^{-1} for amide III (C-N and N-H functionalities), in FTIR spectra. Also, the positions of these bands conform the protein materials, such as 1650 cm^{-1} (random coil) and 1630 cm^{-1} (β-sheet) for amide I, 1540 cm^{-1} (random coil) and 1520 cm^{-1} (β-sheet) for amide II, and 1270 cm^{-1} (β-sheet) and 1230 cm^{-1} (random coil) for amide III. Sericin extracted from fabric shows absorption between 1600 cm^{-1} and 1700 cm^{-1}, confirming amide I absorption which arises predominantly from the C=O stretching vibration and is most useful for determining proteins secondary structure [37]. The peak of sericin at 1540 cm^{-1} belongs to amide II which arises because of the random coil structure. Signature peak for sericin at 1400 cm^{-1} is observed in case of both samples. In addition, the amide III characteristic peak, which arises mainly from the C-N stretching vibration coupled to the N-H plane bending vibration is found to shift in the range of 1240 cm^{-1} to 1250 cm^{-1} corresponding to a change from random coil conformation to β-sheet structure. There is no major variation seen in IR spectra of standard and test samples extracted from various sources and prepared from various techniques. Some researchers have reported similar results [30, 31].

15.9 Fluorescence spectra

The standard and test samples have been recorded for fluorescence spectra in order to determine if the extraction factors influence the conformational state of sericin in any way. Figure 15.3(a) exhibits the spectra. The standard sericin (A) shows a distinct peak at 355 nm, whereas sericin from fabric (B) exhibits a minor peak at around 355 nm and a distinct peak at 380 nm [37]. The fluorescence of a folded protein is a mixture of the fluorescence from individual aromatic residues. Most of the intrinsic fluorescence emissions of a folded protein are due to excitation of tryptophan residues, with some emissions due to tyrosine and phenylalanine.

Typically, tryptophan has the excitation wavelength of 295 nm and an emission peak that is solvatochromic ranging from 300 nm to 350 nm depending on the polarity of the local environment [32]. In Figure 15 3(a), the extracted sericin sample shows a fluorescence pattern which is different from that of the standard sample. This indicates that some changes do take place in the sericin during extraction processes that affect the conformational state of sericin protein. In order to know the relevance of the change in structure, it is necessary to establish the influence of these changes on the biological and performance properties like cell regeneration, antioxidant property and moisture absorption.

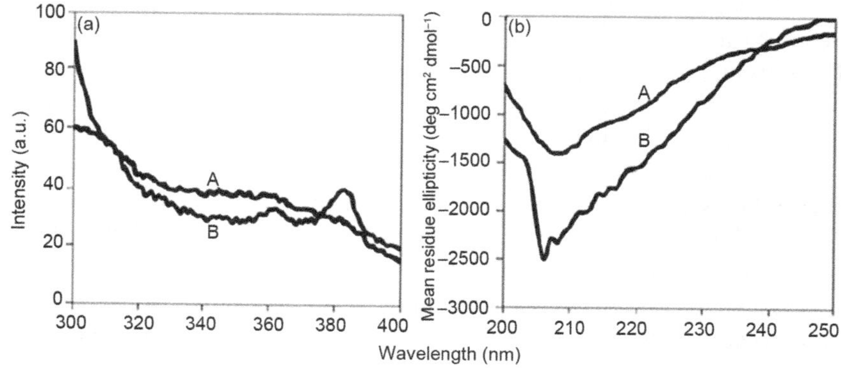

Figure 15.3 (a) Fluorescence spectra, and (b) CD spectra of sericin samples [37]
A – Standard sericin , B – Sericin from silk fabric

15.10 Circular dichroism (CD) spectra

The differences in the left-handed polarized light absorption and right-handed polarized light is measured by CD spectroscopy, and can be used to determine the proteins secondary structure in the far UV spectral region (190–250 nm). At these wavelengths, a chromophore is a peptide bond, and a signal is arises when it is located in a regular, folded environment, α-helix, β-sheet and random coil structures that provide characteristic shape and magnitude of a CD spectrum. The negative peak at 208 nm is characteristic of an α-helix protein and a negative peak at 214 nm is characteristic of β-sheet protein [27]. CD curves (Figure 15.3b) of the IR extracted sericin sample in this study show a negative band at 206–208 nm suggesting α-helix conformation. Gulrajani et al. found that sericin recovered from HTHP degumming liquor shows a negative band at 198 nm and a weak band at 218 nm suggesting random coil and β-sheet configuration, respectively [18]. On the other hand, sericin recovered from alkaline degumming shows a negative peak at 201 nm and 216 nm, revealing the presence of α-helix structure [18]. Similar findings have been reported for the secondary structure of sericin prepared by etanol precipitation. Identical peaks of CD spectra are exhibited by low molecular weight sericin recovered from A. Mylitta [33]. It is possible to conclude from these observations that sericin can exhibit both α-helix and β-sheet structure based on the method of extraction.

15.11 Molecular weight

SDS-PAGE technique has been used to determine the molecular weight distribution of the different sericin samples. Figure 15.4 depicts the results. The sericin from fabric (D) exhibits a diffused band in the molecular weight 65–205 kDa, while standard sericin (B&C) exhibits bands in the low molecule weight range 3.5–43 kDa [37]. The molecular weight of standard sericin is found to be much lower than that of extracted silk. As the two samples are dissolved in water, their solubility also differs. Standard sericin having lower molecular mass (<50 kDa) is soluble in cold water, and extracted sericin having a higher molecular mass between 50 kDa and 200 kDa dissolves only in hot water (90°C).

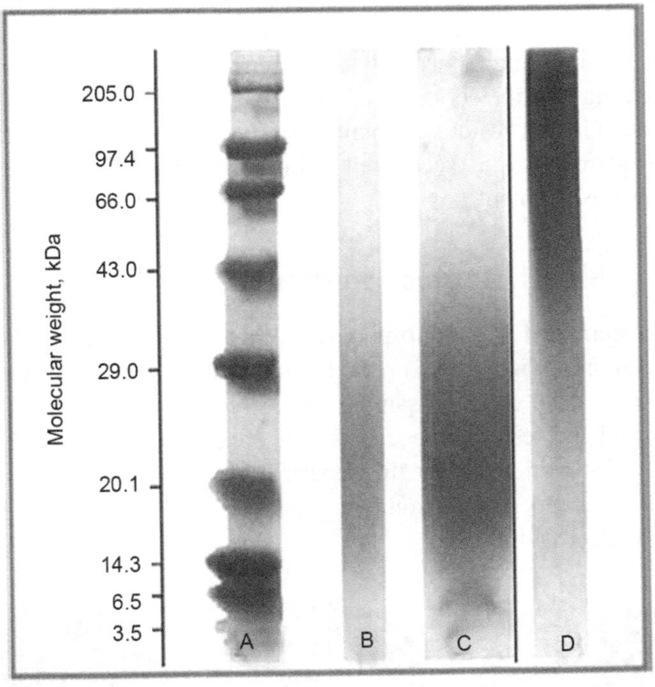

Figure 15.4 SDS PAGE analysis of sericin samples [37]
A – Marker, B – Standard, 20 mg/mL,
C – Standard, 60 mg/mL, and D – Sericin from fabric, 10 mg/mL

This large difference in molecular weight can have a significant effect on the properties and applications of sericin. It has been reported that sericin consists of a group of protein molecules of molecular weight ranging from 20 kDa to 400 kDa. The molecular mass of sericin has been assessed as 309, 177,

145, 134 and 80 kDa, while it is reported at least 15 different polypeptides having molecular weights ranging from about 20 kDa to 200 kDa in the anterior portion of middle silk gland [34, 35]. It has been proved that sericin appears in a continuous distribution between 97 kDa and 14 kDa, while some bands are observed above and below 97 kDa and 14 kDa, respectively.

15.12 Thermogravimmetric analysis (TGA)

TGA has been used to study the thermal behavior of sericin samples. The weight loss trace shows that the initial weight loss is due to the water evaporation. Subsequently, a drastic reduction in weight is detected in the wide temperature range from 220°C upward for both the samples [37]. Similar weight loss pattern has been reported for sericin by Tsukada [36]. It is clear from the weight loss pattern in the thermograms that sericin extracted from the fabric exhibits higher weight loss as compared to the standard sericin, thus indicating that the sericin prepared from fabric is relatively unstable to temperature. Greater residue is obtained in standard sericin powder (9%) than that of the sericin powder recovered from fabric (5%). The results agree well with percent ash content.

15.13 Morphological studies

SEM has been used in the morphological investigation of sericin powders. The micrographs (Figure 15.5) reveal that the particle shape differs much in both the cases. The standard sample comprises of long planar standard size particles, and the test sample consists of spherical globules [37]. Another significant observation is that the spherical globules have a very wide size distribution and appear to be highly aggregated. Such particles generally have problems with solubility.

(a) (b) (c)

Figure 15.5 SEM photographs of sericin samples (x5K magnification) [37]
A – Standard, B – Spray dried sample, and C – Freeze dried sample

To study if the particle shape is indeed characteristic of a drying technique, another test was conducted where the extracted liquor was freeze dried instead of spray drying. Figure 15.5(c) shows the SEM of the powder obtained by freeze drying. Planar-shaped particles are obtained by indicating that the particle shape is a function of the drying technique. All other properties of the new powder are seen to be same as the globular particles.

References

1. Teli MD, and Rane VM. Fibres & Textiles, Eastern Europe, 19(2011), 10.
2. Freddi G, Mosotti R, and Innocenti R. Journal of Biotechnology, 106(2003), 101.
3. Chopra S, and Gulrajani ML. Indian Journal of Fibres and Textile Research, 19(1994), 76.
4. Cho KY, Moon JY, Lee YW, Lee KG, Yeo JH, Kweon HY, Kim KH, and Cho CS. International Journal of Biology and Macromolecules, 32(2003), 36.
5. Yun H, Oh H, Kima MK, Kwak HW, Lee JY, UM IC, Vootlad SK, and Lee HK. International Journal of Biology and Macromolecules, 52(2013), 59.
6. Sarovart S, Sudatis B, Meesilpa P, Grady BP, and Magaraphan R. Reviews in Advanced Material Science, 5(2003), 193.
7. Aramwit P, Dammrongsakkul S, and Srichana T. Biotechnology applications in biochemistry, 55(2010)91.
8. Sasaki M, Yamada H, and Kato N. Nutritional Research, 20(2000), 1505.
9. Ki CS, and Park YH. Macromolecule Research, 17(2009), 935.
10. Gulrajani ML, Gupta SV, Gupta A and Suri M. Indian Journal of Fibre and Textile Research, 21(1996), 270.
11. Bianchi AS, and Colonna GM. Melliand Textilbert, 73(1992), 68.
12. Rangam R, Wang L, Kanwar JR, and Wang X. Journal of Applied Polymer Science, 119(2011), 1339.
13. Gulrajani ML, Sethi S and Gupta S. Journal of Society of Dyers Colour, 108(1992), 79.
14. Gulrajani ML, and Sinha S. Journal of Society of Dyers Colour, 109(1993), 256.
15. Sarma M, Gogoi S, Devi D, and Goswami B. Journal of Scientific and Industrial Research, 71(2012), 270.
16. Capar G, Aygun SS, and Gecit MR, Journal of Membrane Science, 325(2008), 920.
17. Gen CG, Byraktar O, and Basal G, Textil ve Konfeksiyon, 19(4)(2009), 273.
18. Gulrajani ML, Purwar R, Prasad RK, and Joshi M. Journal of Applied Polymer Science, 113(2009), 2796.
19. Vaithanomsat P, and Kitpreechavanich V. Separation and Purification Technology, 59(2008), 129.

20. Kurioka A, Kurioka F, and Yamazaki M. Bioscience Biotechnology Biochemistry, 68(4)(2004), 774.
21. Mandal BB, Ghosh B, and Kundu SC. International Journal of Biology and Macromolecules, 49(2011), 125.
22. Krishnamurthy K, Khurana HK, Jun S, Irudayaraj J, and Demicri A. Comprehensive Review on Food Science and Food Safety, 7(2008), 1.
23. Gupta D, Agrawal A, Chaudhary H, Gupta C, and Gulrajani M. Journal of Cleaner Production, 52(2013)488.
24. Boucau J. Enzymatic and structural characterization linked to mycobacterium tuberculosis pathogenically of proteins, PhD Thesis, The University of Toledo, (2008), 39.
25. Official method of analysis of AOAC international, 17th edition (Association of official analytical chemists, Gaithersburg, MD) 2000.
26. Li G, Liu H, Li T, and Wang. Journal of Material science and Engineering Chemistry, 32(2012), 627.
27. Aramwit P, Siritientong T, and Srichana T. Waste Management Resources, 30(3) (2012), 217.
28. Wu JH, Wang Z, and Xu SY. Food Chemistry, 103(2007), 1255.
29. Lamoolphak W, Eknamkul WD, and Shotipruk A. Bioresource Technology, 99(2008), 7678.
30. Teramoto H, and Miyazawa M. Biomacromolecules, 6(2005), 2049.
31. Zhang XM, and Wyeth P. Science China, 63(2010), 626.
32. Lacowicz JR. Protein fluorescence in topics in fluorescence spectroscopy, Volume 6 (Kluwer Academic Publishers, New York) 2002.
33. Dash R, Ghosh SK, Kaplan DL, and Kundu SC. Comparative Biochemical physiology, Part B, 147(2007)129.
34. Gamo T, Inokuchi, and Laufer H. Insect Biochemistry, 7(1977), 285.
35. Sperague K. Biochemistry, 14(1975), 925.
36. Tsukada M. Journal of Applied Polymer Science, 22(1978), 543.
37. Deepti G, Anjali A, and Abhilasha R. Extraction and characterization of silk sercin, Indian Journal of Fibres and Textile Research, 39(2014), 364.

16

Optimization of technological parameters for steeping silk on small reels

Summary : The properties of silk slices do not exhibit significant difference between steeping on small reels and conventional steeping. The new process of steeping silk slices on small reels can largely satisfy the technological requirements of steeping. The properties of knitted fabric produced by the new steeping as well as conventional steeping methods are found to be similar to each other. But, some special properties relating to the new method of steeping, like abrasion resistance, bursting strength and draping, are better than those of the conventional steeping method. The new method of steeping gives lower cost and higher efficiency in comparison with the conventional steeping of raw silk.

16.1 Introduction

Steeping involves immersing skein silk into aqueous solution with agents that cause swelling and softening of sericin, thereby rendering the raw silk with good performance properties, like compliance, flexibility, lubricity, etc. It has suitability for subsequent processes and also the quality of the product is assured [1, 2]. At present, for the purpose of obtaining skein silk, the following processes are necessary: balance of silk slices on small reels; the vacuum permeation treatment of silk slices on small reels; balance of wet silk slices on small reels; re-reeling balance of silk slices on large reels; lacing and checking balance of silk slices; hank of silk; weighing of silk; color blending; packing batch of products [3]. All of the mentioned processes are accomplished in a silk factory. In a knitting factory, in order to meet the process requirements of winding, twisting, knitting and so on, it is necessary to treat skein silk by the processes of steeping raw silk, oven drying and airing. These processes need a large amount of manpower, material resources and lot of space for airing. Hence, the above-mentioned conventional process can be omitted if the skein silk meets the requirements of winding, twisting and knitting through adding a steeping assistant during the processes before re-reeling in the silk factory. This may improve production efficiency and reduce production cost. Therefore, it is necessary to find a suitable process. In this chapter, a new steeping method is developed in which steeping assistant is added into the solution during the vacuum permeation treatment of silk slices on small reels, namely, steeping silk slices on small reels. Orthogonal array

DOI: 10.1201/9781003735427-16

designs have been widely used with the advantage of reducing the number of the tests while giving reliable results to the optimize test [4–7]. This chapter aims at evaluating the experimental conditions that provide the excellent performance of raw silk based on orthogonal array designs for steeping silk slices on small reels. Three parameters, namely, time of balance drying, concentration of assistant and vacuum number, are studied and optimized by a four-level orthogonal array design, with breaking strength, breaking elongation, compliance, loop elongation and loop strength as the performance index. As a kind of natural protein fiber, the structure and performance of silk may be affected by different processing methods, which may further affect the quality of silk and fabric [8]. So it is necessary to compare silk slices on small reels after steeping and skein after conventional steeping, as well as raw silk without steeping in terms of quality and micro-structure of the silk. Also, the knitted fabric produced by the new steeping method has been compared with that of the conventional steeping method.

16.2 Technical details

For the study, raw silk conforming to Grade 6 A silk slices on small reels (specification: 20/22 D, namely 22.2/24.4 dtex) have been chosen. Single jersey knitted fabrics have been produced and scoured. The balance drying time, assistant concentration and vacuum number, which effect properties of silk slices on small reels, are studied. The method of orthogonal array L16 (43) is used to determine the optimal technological parameters. Some other experimental parameters, such as degree of vacuum, temperature of water, bath ratio and the assistant applied to the steeping weft, are selected as 0.08 MP, 40 ± 1°C, 1:100 and EKL-101, respectively [18]. The experimental parameters of conventional weft steeping are 40 ± 1°C, 1:100, EKL- 101 and 3h for water temperature, bath ratio, the assistant applied to steeping weft and steeping time, respectively. A three-factor, four-level factorial design is used to evaluate the following factors effect on properties of raw silk: time of balance drying (A), concentration of assistant (B) and vacuum number (C). In order to estimate the best condition for steeping silk slices on small reels, 16 experiments are performed. Yarn tensile strength tester is used to test the breaking strength, breaking elongation, loop strength, loop elongation and elasticity modulus of raw silk. The test condition is as follows: the active length is 500 mm, the drawing speed is 500 mm/min and the original load is (0.05 ± 0.01) cN/dtex. All mechanical tests should be carried out at the temperature of (20 ± 2.0)°C and relative humidity of (65.0 ± 4.0)%. These samples should be placed under the above condition for 12 h before tests. The elasticity modulus reflects the compliance of raw silk, which is inversely proportional to the elasticity modulus. The sericin content has also been tested and silk degummed completely. After gold coating, a scanning electron microscope is utilized to obtain the morphology of the raw silk with 5 kV acceleration voltage and 1000 × magnification. X-ray diffraction (XRD) measurements are performed with an X-ray diffractometer equipped with a Ni-filtered CuKa radiation source at 40 kV voltage and 40 mA tube current. The diffraction intensities are measured from 2μ = 5° to 50° at a rate of 2μ = 2 mm/min. The crystallinity of raw silk has been calculated [10, 11]. The properties tested in the knitted fabrics include tensile properties, bursting strength, abrasion resistance, and draping.

16.3 Orthogonal array analysis

Repeated test is done without the blank column in the orthogonal array design to get reliable results [12]. Twenty samples are tested for every performance index. Initial data are omitted. The orthogonal array, date analysis and analysis of variance (ANOVA) are obtained using SPSS software [13, 14]. The ANOVA indicates that concentration of assistant (B) and time of balance drying (A) play important roles in the properties of raw silk. By comparison, we find that the effect of vacuum number (C) is less important than the other two factors. Based on the experimental assignment, the mean of property for the corresponding factors at each level has been calculated.

16.4 Influence of duration of balance drying on properties of raw silk

It has been found that the different durations of balance drying (drying for 0, 2, 4 and 6 h) influence the breaking strength, breaking elongation, loop strength, loop elongation and elasticity modulus. When drying time is 2 h, breaking strength, breaking elongation, loop strength and loop elongation of raw silk reach a maximum, and the elasticity modulus of raw silk reaches a minimum at drying for 4 h [18]. Drying time of 2 hours has been chosen for drying as the optimum level depending on an overall consideration.

16.5 Influence of concentration of assistant on raw silk properties

The breaking strength, breaking elongation, loop strength, loop elongation and elasticity modulus is found to be affected by the different assistant concentrations (0.75%, 1.00%, 1.25% and 1.50%) [18]. When the concentration of assistant is 1.25%, breaking strength, breaking elongation, loop strength and loop elongation of raw silk reach a maximum, and the elasticity modulus of raw silk reaches a minimum at the assistant concentration of 1.50%. A concentration of the assistant for 1.25% has been chosen as the optimum level.

16.6 Influence of vacuum number on raw silk properties

The breaking strength, breaking elongation, loop strength, loop elongation and elasticity modulus are affected by the different vacuum numbers (5 times, 10 times, 15 times and 20 times). As to breaking strength and loop strength, the values reach a maximum when the vacuum number is 15 times. As to breaking elongation and loop elongation, the value reaches a maximum when the vacuum number is 5 times and 10 times, respectively. At the level of 5 times for the vacuum number, the elasticity modulus of raw silk reaches a minimum. According to these results, it is difficult to determine the optimum level for vacuum number. Based on the above ANOVA calculations, it seems that the effect of vacuum number is less important than that of the other two factors [18]. That is, vacuum number (C) is a secondary factor. Therefore, we can select the shortest vacuum number as 5 times taking other factors into account, such as production efficiency, convenience of operation and so on. The technological parameters of steeping conditions for silk slices on small reels after optimization are determined as 2 h for balance drying, 1.25% for assistant concentration and 5 times for vacuum number, so as to achieve good properties of raw silk.

16.7 Comparison of silk properties

The breaking strength, breaking elongation, loop strength, loop elongation, elasticity modulus and sericin content of skein after conventional steeping (a), silk slices on small reels after steeping (b) and raw silk without steeping (c) have been compared.

16.8 Comparison of sericin content

Sericin content has been calculated and found to have no significant difference in the sericin content of the three samples. In addition, sericin content of samples a and b is slightly greater than sample c, namely sericin content is slightly increased after steeping. This may be because strand silk absorbs a certain amount of oil after steeping [15]. There is no significant difference in CV. It has been observed that the strength of samples a and b is slightly lower than sample c, namely, the strength of samples a and b is slightly decreased after steeping [18]. This may be because the molecules are farther apart from each other, and intermolecular interactions decreases. There is no significant difference between samples a and b. Referring to GB/T 1797-2008 raw silk, we find that all breaking strength remain at the 6 A grade, which is the highest silk level. The CV of the three samples is similar. As to elasticity moduli, we can see those of samples a and b are slightly lower than sample c, namely, the compliance of sample c is improved after steeping. This may be because water molecules enter the amorphous of fibroin, cracking intermolecular secondary bonds and making the molecular chain slip [16]. In addition, compliance of sample b is slightly lower than that of sample a. It has been found that both breaking elongation and loop elongation of sample b are the lowest among the three samples.

16.9 Comparison of raw silk structure

16.9.1 Analysis of surface morphology

Figure 16.1 depicts the surface morphologies of raw silk observed under SEM. Raw silk possesses several fibrils that are adhered to each other by sericin. The tiny particles on the surface of raw silk, which may be sericin tumors, lead to the rough surface of raw silk [18]. The obvious differences are hard to differentiate from each other, which implies that the raw silk surface morphology is unaffected by steeping silk slices on small reels.

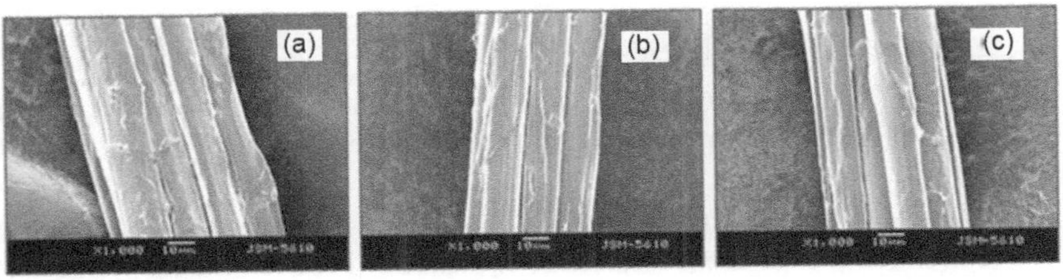

Figure 16.1 SEM images of raw silk [18] (a) Skein after conventional steeping, (b) Silk slices on small reels after steeping (c) Raw silk without steeping

16.9.2 Analysis of X-ray diffraction (XRD)

There are no obvious differences among the three curves. While it is difficult to observe the diffraction peak from these curves, the PeakFit program is used to separate the overlapping bands. It is clear to see that the main diffraction peak appears at around $2\mu = 10.8°, 18.3°, 20.6°, 24.4°, 28.3°$. With the PeakFit program, the crystallinity of raw silk has been calculated. It has been found that there are slight difference for the crystallinity of raw silk and no obvious difference in the structure of the three samples. The lower crystallinity of sericin on the raw silk surface leads to an effect on the crystallinity of silk [17]. Also, the three samples have been compared relatively for crystallinity [18].

16.10 Comparison of knitted fabric properties

After scouring, the knitted fabrics produced from two samples have been compared for properties. It has been found that the properties of the two knitted fabrics are very close. In addition, the value of abrasion resistance and bursting strength of knitted fabric 2 is greater than that of knitted fabric 1. The drape coefficient of knitted fabric 2 is lower than that of knitted fabric 1 [18]. The abrasion resistance, bursting strength and draping of the knitted fabric produced from silk treated by the new steeping method is better than that treated by the conventional steeping method. The new steeping method is found to be feasible as can be concluded from the studies.

References

1. Lou GX. Soft process of silk used for knitting and composite yarn. SILK 1996; 8: 26–27 (in Chinese).
2. Guo XN. Research on silk steeping assistants TWO1. J Zhejiang Silk Eng 1984; 1: 16–21 (in Chinese).
3. Wang XY. The new technology of silk processing. Beijing: China Textile & Apparel Press, 2001, p.189 (in Chinese).
4. Bakhtiari MR, Faezi MG, Fallahpour M, et al. Medium optimization by orthogonal array designs for urease production by Aspergillus niger PTCC5011. Process Biochem 2006; 41: 547–551.
5. Yamini Y, Saleh A and Khajeh M. Orthogonal array design for the optimization of supercritical carbon dioxide extraction of platinum (IV) and rhenium (VII) form a solid matrix using cyanex 301. Sep Purif Technol 2008; 61: 109–114.
6. Chang Y, Liu B and Shen B. Orthogonal array design for the optimization of supercritical fluid extraction of baicalin from roots of Scutellaria baicalensis Georgi. J Sep Sci 2007; 30: 1568–1574.
7. Park CK and Ha JY. A process for optimizing sewing conditions to minimize seam pucker using the Taguchi method. Textil Res J 2005; 75: 245–252.
8. Wang R, Jiang WB, Li S, et al. Application research on infrared drying in silk re-reeling process. Textil Res J 2012; 82: 1329–1336.
9. Shan XH and Xu H. Study on the technological process of soaking in silk knitting. J Xinjiang Univ 2002; 1: 123–127 (in Chinese).

10. Ling S, Qi Z, Knight DP, et al. Synchrotron FTIR microspectroscopy of single natural silk fibers. Biomacromolecules 2011; 12: 3344–3349.

11. Cheng YG, Lin H and Chen YY. Research on the state of aggregation structure of Mulberry silk fiber treated with Nano-ZnO. Jiangsu Textil 2008; 3: 43–53 (in Chinese).

12. Deng ZW, Yu P and Chen L. Application of SPSS software in orthogonal design and result analysis. Comput Stud 2009; 5: 15–17 (in Chinese).

13. Wang Y, Harrison M and Clark BJ. Experimental design for a basic mixture on a fluorinated packing: the effect of composition of the mobile phase. J Chromatogr A 2006; 1105: 77–86.

14. Zheng CG, Wang MM, Wang YL, et al. Optimization of biosurfactant-mediated oil extraction form oil sludge. Bioresour Technol 2012; 110: 338–342.

15. Yu T, Li MS and Guo XN. Initial study of laboratory measurement of the result of raw silk after steeping. Zhejiang Silk Sci Technol 1981; 1: 21–25 (in Chinese).

16. Chen JY and Yao YF. On the steeping technique of raw silk for knitting. J Zhejiang Silk Eng 1990; 7: 1–5 (in Chinese).

17. Zhou ZS. A study on the crystallinity of sericin by X-ray diffraction. J Soochow Silk Eng 1982; 2: 28–40 (in Chinese). 2218 Textile Research Journal 83(20).

18. Xiaoxiao Wang, Yaquin Fu, Chengxia Liu, Dong dong Dai, and Wenbin Jiang, Technological parameters optimization by orthogonal array designs for steeping silk slices on small reels, Textile research journal, 2013, 83(20)2211.

17

Drying of silk in re-reeling process by infra red technique

Summary : A medium-wave IR drying has been used as an alternative to conventional steam drying of silk slices. For the drying of dry silk slices, medium wave infra red drying has been used in the silk re-reeling process as an option over conventional steam drying. The technique has no negative influence on the formation of silk slices, the quality, structure and properties of raw silk, and can also be applied for infrared drying in silk re-reeling process. Compared with conventional steam drying, the temperature inside the re-reeling machine compartments using medium-wave IR drying can be lowered more than 10°C, which reduces the heat transmission from the compartments to the surrounding environment. Moreover, it enhances the working environment and workers' efficiency. The energy consumption is economical, as the energy consumption required by IR drying is less than one-sixth of that required by conventional steam drying.

17.1 Introduction

Re-reeling involves the rewinding of a small reel of raw silk and its conversion onto a big reel or a bobbin. It gives raw silk with the necessary properties like flexibility, strength, elongation, a certain degree of dryness, etc. In order to attain a good quality raw silk, silk slices are dried during the re-reeling process [1]. Hence, drying is considered to be crucial in the silk re-reeling process. At present, the steam pipe drying method is widely adopted by many reeling factories, along with hot air drying. For conventional steam pipe drying, silk slices are dried via hot air heated by the steam pipes, which are set in a re-reeling machine and filled with vapor generated by a coal-fired boiler and transported through the pipelines. However, this method has many disadvantages, such as uneven heating of silk slices, high-energy consumption, low-thermal efficiency, high quantity of sulfur dioxide emissions, serious environmental pollution, uncontrollable heating temperature, and poor working environment for workers because of the high heat loss. In contrast with steam pipe drying, hot air drying has enhanced the drying rate and uniformity of products. However, this method also has some shortcomings, such as high heat consumption and heat loss, large space requirement, high equipment costs, and so on [2]. Obviously, the method does not meet the requirements of a low-carbon economy, which advocates low consumption, low emissions, and low

DOI: 10.1201/9781003735427-17

pollution. Therefore, it is necessary to find a new raw silk drying method with high efficiency, high quality, and low cost. Infrared (IR) heating transfers heat in the form of electromagnetic waves and generates heat in the material center directly without heating the surrounding air. Thus, IR heating is recognized as a drying method with high thermal efficiency, fast heating rate, simple equipment requirements, easy control of heating temperature, convenient application, significant savings of operation space and energy consumption [3,4]. In recent years, some studies on infrared drying applied in the plastic and food products industry have shown that this drying method does not affect or even improve the drying quality of products [5–7]. Early in the 1940s, drying of textiles by infrared radiation was reported, and then the research into the application of infrared technology for the pre-drying of polyester, cotton, acetate, etc., was carried out [8–10]. Langlois and Maisonneuve presented conclusions that high energetic efficiency, low installation costs and space requirement of IR drying gave it a decided advantage through the study on IR drying of viscose, wool, acetate and nylon, etc. [11]. However, there is no research about IR technology utilized in silk drying during the re-reeling process. For steam drying, the most widely used method in the silk re-reeling process, silk slices have to be dried by huge quantities of hot drying air, which results in large energy consumption, high temperature of compartments and poor working environment. Therefore, it cannot only reduce the costs of drying but also helps to improve the working environment. Silk is a kind of natural protein fiber. Various heating techniques in the drying process can influence the structure and performance of silk proteins, and in turn influence the raw silk quality. Comparison has been made with regard to the quality, micro-structure of silk slices, and energy consumption for IR drying in the silk re-reeling process with conventional steam heating.

17.2 Technical details

Silk slices on small reels (specification, 20/22 D) have been taken from an automatic reeling machine. The IR re-reeling machine (Figure 17.1), incorporating the re-reeling machine, IR radiation source, temperature sensor, proportional integral derivative (PID) controller, electric energy meter for energy consumption measurement, and other components, have been specially designed for the purpose [22]. There are two medium wave infrared heating tubes in the heating component in the re-reeling machine. At the base of the re-reeling machine compartment, a parabolic polished aluminum reflector has been fixed. Compared with far IR, medium-wave IR penetrates an object deeper, and thus can be used in improving the quality of silk slices. Hence, medium-wave IR heating tubes have been used.

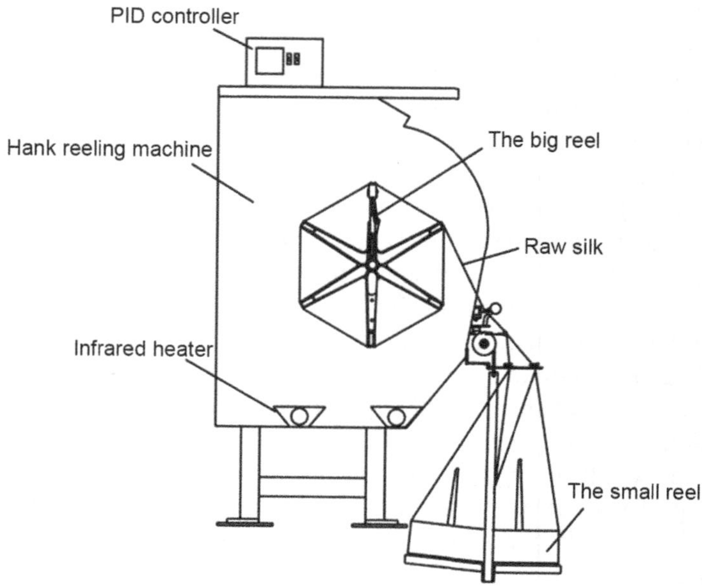

Figure 17.1 IR reeling machine (experimental) [22]

17.3 Drying of re-reeled silk

The silk slices after humidification and equilibration have been rewound on small reels and divided into two parts, of which one part was dried with a factory re-reeling machine by conventional steam drying, and the other was dried using the newly designed re-reeling machine through IR drying. The newly designed re-reeling machine can re-reel and dry five silk slices at a time. The five silk slices were numbered, 1 to 5 in sequence. No. 1 and No. 5 were situated at the opposite ends of the machine compartment, whereas No. 3 was placed at the center of the compartment. No. 1 and No. 5 were equidistant from the rest of the silk slices, whereas No. 3 was the farthest from the first two [22]. Hence, No. 1 was compared with No. 3 to determine the effect of different compartment locations on IR drying. The steam-dried silk slices were used as reference samples and labeled 'S'. The randomly selected raw silk from the outer and inner layers of the silk slices were marked 'a' and 'b', respectively (i.e. 1-a for the outer layer of IR-dried silk No.1, 1-b for the inner layer of silk No. 1, S-a for the outer layer of steam-dried silk No. S, and so on) [22]. The experiment was performed as follows: first, the IR re-reeling machine was switched on; second, the initial heating temperature of the IR heating tubes was set using the PID controller; third, the temperature switch was opened and the IR heating tubes began to heat; fourth, the compartment temperature reached the desired level and was kept stable; and finally, the silk slices were re-reeled and dried. Based on earlier trials, the temperature of the compartment inside the IR re-reeling machine has been maintained at 30–35°C so as to achieve well-formed silk slices.

17.4 Determination of raw silk performance

17.4.1 Determination of moisture regain of silk slices after doffing

The doffed silk slices of the big reel have been evaluated for moisture regain after re-reeling adopting the following equation [12]:

$$R = \frac{w - w_1}{w_1} \times 100\%$$

where w and w_1 are the weights of the sample before and after oven-drying, respectively, and R is the moisture regain.

17.5 Quality inspection of raw silk

The raw silk quality has been evaluated with regard to clearness, neatness, winding, cohesion, breaking strength, and breaking elongation rate, which may be influenced by the drying technique. These factors have been inspected by the State Quality Supervision and Inspection Center of cocoon silk as per specified needs. Clearness and neatness tests were performed to determine the types, number, and distribution of big and small defects on the silk fiber against photographic comparison standards [22]. The tests were performed on a seriplane in a dark room with light fixture. Each sample was wound at a pre-determined pitch (80 lines per 25.4 mm) onto a quadrate blackboard (1359 × 463 × 37 mm) in 10 equal parts and the width of each part was 127 mm. Soft light shone on the blackboard evenly with average illumination of 400 lx. The winding test was conducted by examining the winding break times using a silk winder under a specified recoiling tension, coiling speed, and time.

According to the requirement of the existing test method for raw silk, the coiling speed was 165 m/min, the readiness time and the inspection time were 5 min and 120 min, respectively. A piece of raw silk is reeled by several cocoon filament fibers that adhere to one another via their sericin.

Raw silk can withstand all kinds of friction during processing because of the adhesive property of sericin. Raw silk cohesion refers to the friction resistance of the silk filament without generating splitting. The cohesion test was carried out on a Y731 cohesion tester. Silk fibers were wound to 10 hooks on both sides of the machine framework and the different parts of silk filaments were rubbed under uniform and constant tension with a rubbing speed of 130 times/min at the same time. The breaking strength, breaking elongation of B. mori silk filaments (steam-dried and IR-dried) were determined using YG065 electronic multifilament tensile strength tester. The active length was 100 mm and drawing speed was 150 mm/min. The original load calculated by 0.5 cN/dtex and the fixed elongation was 5%. The results were obtained by averaging 10 samples. The winding, cohesion, breaking strength and breaking elongation tests were carried out at laboratory atmosphere, 25°C and 65% relative humidity and the samples were placed under these conditions for more than 12 h before measurement.

17.6 Scanning electron microscopy observation

The morphologies of raw silk from the outer and inner layers of both IR- and steam-dried silk slices have been viewed longitudinally by means of SEM at 5 kV acceleration voltage.

17.7 X-ray diffraction analysis

X-ray diffraction system has been used to analyze the IR-dried and steam-dried raw silk in ground and powdered form, and crystallinity of raw silk calculated [13, 22].

17.8 Influence of IR drying on silk properties

17.8.1 Moisture regain of big reel silk slices after doffing

The moisture regain of IR-dried and steam-dried silk slices after doffing has been determined. The obtained R values are 8.39% and 8.54%, respectively, which are almost the same [22]. Therefore, both silk slices meet the 7.5%–9% requirement for big reels doffed from the re-reeling machine [14].

17.9 Raw silk quality

The quality of raw silk is closely related to the weaving technology, fabric quality, production cost, trade value, and market competitiveness [15]. Hence, the raw silk quality after IR drying is considered crucial for evaluation of the feasibility of the drying method.

17.10 Neatness and clearness of raw silk

The clearness and neatness of the samples have been determined [22]. There is no significant difference in the clearness and neatness of raw silk dried using the two methods, indicating that the above-mentioned qualities of silk are not affected by IR drying and also its performance is same as that of steam drying.

17.11 Winding, cohesion, and mechanical properties of raw silk

The winding, cohesion, breaking strength, and breaking elongation rates of the IR-dried and steam-dried raw silk have been determined. The cohesion and breaking strength values of the two groups of silk are very close, whereas the breaking elongation rate of the IR-dried silk is slightly decreased [22]. However, all qualities remain at the highest silk level, the 6A grade. In addition, the requirements for weaving and knitting are met, indicating that IR drying has no significant effect on the cohesion and mechanical properties of raw silk. No. 1 and No. 3 also show no obvious difference between their performances, suggesting that the performances of silk situated at the opposite ends and at the center of the IR re-reeling machine are stable. Hence, the raw silk quality is not influenced by IR drying in comparison with conventional steam drying.

17.12 Influence of IR drying on raw silk structure

17.12.1 Analysis of surface morphology

The longitudinal surface morphology of the outer and inner layers of the IR-dried and steam-dried silk are explained. Raw silk possesses uneven grooves on the longitudinal surface, and the fibrils are regularly arranged, roughly parallel with the fiber axis, and adhered to each other via sericin. In the absence of degumming, the surface of raw silk is rough and has numerous tiny particles, which may be sericin tumors [22]. No obvious difference is observed in the SEM images of each layer of raw silk dried using the two methods. The surface morphologies of the IR-dried No. 1 and No. 3 silk layers show no significant difference. Thus, IR drying has no effect on the surface morphology of raw silk compared with steam drying.

17.12.2 XRD analysis

There is a near similarity in the XRD curves for the medium-wave IR-dried and steam-dried raw silk. All curves exhibit a strong diffraction peak at around 2μ = 20.6°, suggesting that the main structure of the silk fiber has not changed. The other major diffraction peaks of the six samples are at around 10.7°, 18.3°, 24.2°, and 27.9°, corresponding to the characteristics of Bombyx mori silk fiber with a highly oriented silk II (b) structure, which has few migrations compared with that of pure silk fibroin [16, 22]. This result can be attributed to the effect of sericin in raw silk. However, the crystalline structure of raw silk is still silk II-dominated [17]. The crystallinity of raw silk was calculated via the peak separation method using Peakfit v. 4.12. It has been found that medium-wave IR drying has little effect on the crystallinity of raw silk compared with steam drying.

17.13 FT-IR spectral analysis

The FT-IR spectra of raw silk after drying have been analyzed. The spectra of the IR-dried silk layers are almost the same as those of the steam-dried silk. However, their absorption strengths are different [22]. The FT-IR spectra of the six samples show strong absorption bands at 3300 cm^{-1} (N–H stretching), 1654 cm^{-1} (C–O and C–N stretching), and 1515 cm^{-1} and 1230 cm^{-1} (predominantly N–H bending and C–N stretching) [18]. The peak of the silk fiber at 1654 cm^{-1} belongs to the amide I range, corresponding to a typical random coil conformation. The absorption of the silk fiber at 1515 cm^{-1} and 1230 cm^{-1} are attributed to amides I and II, respectively, corresponding to the b-sheet conformation [19, 20]. The medium wave IR drying does not significantly influence the silk fiber surface groups in comparison with steam drying, as indicated by the peak relative intensities of the six samples which are basically consistent.

17.14 Drying temperature and energy consumption analysis

The drying temperature inside the IR re-reeling machine is lower than that of steam drying, since IR heating does not require a heat transfer medium to induce a fast drying rate [22]. In this experiment, the temperature of the compartment inside the IR re-reeling machine was kept about 33°C. Meanwhile, the temperature produced by the steam drying method is around 45°C, which is the same temperature used in factories. The compartments in the re-reeling machine are open-ended. Thus, the lower temperature is an advantage in the working environment. The energy consumption during the re-reeling process was determined by an energy meter. The re-reeling of five pieces of approximately 900 g silk slice require 0.8 kWh, indicating that an energy consumption of 3420 kJ is needed to dry 1 kg of raw silk. A total of 12 kg of 0.155 MPa water vapor is needed to dry 1 kg of re-reeled silk via steam pipe heating [21]. Pipeline transportation losses are excluded, indicating that more than 25,000 kJ should be consumed to dry 1 kg of raw silk at a condensed water temperature of 30°C. The energy consumption of medium-wave IR drying is significantly lower, amounting to less than one-sixth of that required in conventional drying. Hence, medium-wave IR drying increases the economic efficiency, and also significantly decreases the sulfur dioxide emissions produced by burning coal, as well.

References

1. Shigematsu M. Drying of silk slices during rereeling process [J]. Silk Textile Technol Overseas 1983; 2: 9–13, Chinese.
2. Tan ZH. Research on the revolution of heating and drying systems of re-reeling machine. Sichuan Silk 1999; 3: 18–25, Chinese.
3. Krishnamurthy K, Khurana HK, Soojin J, Irudayaraj J and Demirci A. Infrared heating in food processing: An Overview. Comp Rev Food Sci Food Safety 2008; 7(1): 2–13
4. Cassidy D, Ebiana AB, Salem A and Rashidi M Numerical analysis of a radiant drying oven for web applications. Int J AdvManuf Technol 2007; 32(3): 238–256.
5. Vogt M. Infrared drying lowers energy costs and drying times. Plast Add Comp 2007; 9(5): 58–61.
6. Shi J, Pan Z, Mchugh TH, Wood D, Hirschberg E and Olson D. Drying and quality characteristics of fresh and sugar-infused blueberries dried with infrared radiation heating. LWT-Food Sci Technol 2008; 41(10): 1962–1972.
7. Lee SH, Ko SC, Kang SM, Cha SH, Ahn GN, Um BH, et al. Antioxidative effect of Ecklonia cava dried by far infrared radiation drying. Food Sci Biotechnol 2010; 19(1): 129–135.
8. Preston JM and Chen JC. Some aspects of the drying and heating of textiles. J Soc Dyers Colour 1946; 62(12): 361–364.
9. Paul GT and Wilhelm RH. Radiation drying of textiles. Textile Res J 1948; 18(10): 573–597.
10. Broadbent AD, Cote B, Fevteau T, Khatibi-Sarabi P and Therien N. Pre-drying textile fabrics with infrared radiation. Textile Res J 1994; 64(3): 123–129.
11. Langlois C and Maisonneuve R. Electric infrared technology in textile drying processes. In: National Technical Conference of AATCC, Atlantic City, USA, 1986, pp.85–94. Research Triangle Park, NC: AATTC.
12. CSBTS. Determination of moisture content and moisture regain of textile – oven-drying method. Beijing: Standards Press of China, 1997, p.1, Chinese.

13. Hermans PH and Weidinger A. Quantitative investigation of the X-ray diffraction picture of some typical rayon specimens. Textile Res J 1961; 31(6): 558–571.

14. Huang JT, Zhu WM, Xia JG and Xiang ZH. Chinese Silk Pandect. Chengdu: Sichuan Science and Technology Press, 1996, p.750, Chinese.

15. Li T and Lu WM. Introduction to quarantine inspection and textiles import and export inspection. Shanghai: China Textile University Press, 2005, p.160, Chinese.

16. Lu YH, Lin H, Chen YY, Wang C and Hua YR. Structure and performance of Bombyx mori silk modified with nano-TiO_2 and chitosan. Fibers Polym 2007; 8(1): 1–6.

17. Xie RJ, Huang SE, Li CP, Cao JG and Wang B. Effect of non-cocoon cooking on silk fiber. Silk Monthly 2010; 4: 20–23, Chinese.

18. Freddi G, Tsukada M and Beretta S. Structure and physical properties of silk fibroin/polyacrylamide blend films. J Appl Polym Sci 1999; 71(10): 1563–1571.

19. Freddi G, Pessina G and Tsukada M. Swelling and dissolution of silk fibroin (Bombyx mori) in N-methyl morpholine N-oxide. Int J Biol Macro 1999; 24(2–3): 251–263.

20. Lu Q, Hu X, Wang X, Kluge JA, Lu S, Cebe P, et al. Water-insoluble silk films with silk I structure. Acta biomaterialia 2010; 6(4): 1380–1387.

21. Zhejiang Silk Company. Handbook for Silk Reeling, 2nd edn. Hangzhou: China Textile & Apparel Press, 1992, pp.382–383, Chinese.

22. Rui W, Wenbin J, Shan L, Hongbin Y, Yubing D, and Yaquin F. Application research on infrared drying in silk re-reeling process, Textile Research Journal, 82(2012)(13), 1329.

18

Dyeing of silk with bio ecological mordant

Summary : In the natural dyeing of silk, monascorubrin pigment extracted from the solid-state fermentation of monascus offers an option. Silk fabrics have been effectively dyed with monascorubrin pigment using terrae rare mordanting. The dye uptake and color strength values can effectively be enhanced by praseodymium chloride. The simultaneous mordanting technique resulted in the best adsorption capacity. When the pH increased in the range 3–11, the adsorption capacity reduced. The maximum uptake of dye has been at pH value of 3. The dyeing rate of monascorubrin pigment alone was lower than for simultaneous mordanting with praseodymium chloride. The time of half dyeing reduced and the equilibrium dye uptake increased drastically for the mordanted sample. The use of mordant not only improved color strength but also enhanced color-fastness. The covalent bond between the dye molecule, fiber and mordant resulted in the good fastness of the mordanted samples. The monascorubrin pigment showed very good build-up for the silk fabric.

18.1 Introduction

Traditional natural pigments generally had their source from animal, plant or mineral origin [1–3]. Red pigment is obtained from madder, blue pigment from Indigo, and black pigment was obtained mainly from logwood. It has been known that some biological sources like bacteria, fungi and mildew can also be utilized for natural pigment production [4, 5]. In view of their biodegradability and less harmful effect on the environment, the application of natural pigments in textile coloration technology has drawn increasing worldwide attention. It could meet the consumer needs for safe and healthy life [6, 7]. Nevertheless, dyeing with natural pigments has some problems such as low reproducibility, low dye exhaustion and poor fastness. In order to improve the dye uptake and color-fastness, most of the dyeing processes were conducted by adding metal-based mordants [8]. These mordants comprised heavy metal ions such as copper, iron, aluminium, etc., which could cause environmental problems [9, 10]. For these reasons, much effort had been spent to exploit ecological mordants. Terrae rare showed high compatibility with the environment and ecological safety, so it could be used as an eco-mordant for dyeing. Monascorubrin is extracted from the solid state fermentation of monascus. It is a water-soluble pigment with high purity and high color value. The stability of monascorubrin in the pH range 3–11 is good. The stability of

DOI: 10.1201/9781003735427-18

the pigment against heat is also excellent. Monascorubrin has anticarcinogenic, antioxidative and hypolipidemic activities, which is usually used as an excellent food colorant for humans with its associated therapeutic benefits [11, 12]. In this chapter, we investigate the mordant dyeing of monascorubrin on silk fabric. To reduce environmental pollution, terrae rare (praseodymium chloride) was used as a mordant. The influence of various mordanting techniques (pre-mordanting, simultaneous mordanting, and post-mordanting) on the dyeing properties has been studied. The color fastness and kinetic parameters and color-fastness of monascorubrin have also been studied.

18.2 Technical details

Commercially available silk fabric has been used in the investigation. The samples have been scoured and bleached so as to remove contamination and pigment, prior to dyeing. Figure 18.1 shows the chemical structure of monascorubrin.

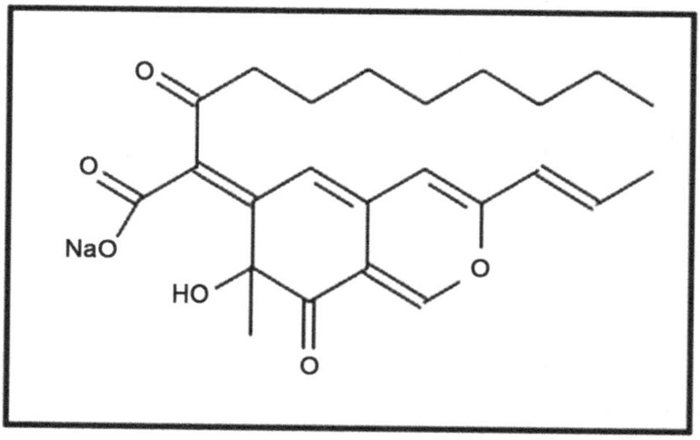

Figure 18.1 Chemical structure of monascorubrin [13]

18.2 Techniques of mordanting

Three techniques of mordanting have been adopted, namely, premordanting, simultaneous mordanting and post-mordanting.

1. In the pre-mordanting, silk fabric has been immersed in 2% o.w.f. in praseodymium chloride solution at 60°C for 1 hour. The sample has then been dried at room temperature.
2. In the simultaneous mordanting, the silk fabric has been introduced into the dye liquor of praseodymium chloride and monascorubrin. The mordant/pigment ratio is chosen as 1:1.

3. In the post-mordanting, the dyed silk fabric has been impregnated with 2% o.w.f. mordant solution at 60°C for 1 hour. The silk fabric has then been washed with water to remove the unfixed dye, and then air-dried at room temperature.

18.3 Procedure of dyeing

An exhaustion technique has been used in the dyeing of the silk fabrics. Dyeing has been done under the following conditions: the bath ratio 1:30, the pH in the bath was 3, the dye concentration was fixed at 2% o.w.f. The baths were prepared at 30°C and heated at 1°C/min till attainment of equilibrium temperature of 80°C. Dyeing was then continued at 80°C for 1 hour. The samples have then been rinsed well after dyeing and dried at ambient temperature.

18.4 Uptake of dye

A spectrophotometer has been used to measure the optical density of the dye solution [13]. Colorimetric measurements have been done on the dye baths and rinsing baths to determine the quantity of unabsorbed dye and this permitted calculation of dye quantity on the fiber. Dye uptake was determined by

$$\text{Dye uptake} = q_2/q_1 \times 100\%$$

where q_1 is the total quantity of pigment in the dyebath, and q_2 is the quantity of pigment adsorbed by fabrics.

18.5 Measurement of color strength

The color strength (K/S value) of the dyed samples has been assessed using the Kubelka Munk equation.

18.6 Color-fastness

The washing-fastness and light fastness has been measured based on Chinese Standard [11,12].

18.7 Influence of mordanting techniques on the dye uptake and color strength (K/S)

The influence of mordanting techniques on the dye uptake and K/S value of silk fabrics has been studied. Compared with other mordanting techniques, the simultaneous mordanted sample has greater depth of shade. Terrae rare mordant could mediate interactions between the pigment and silk fiber [13]. Pigment-terrae rare ion complex would be formed on fiber molecules. The reaction mechanism is explained. The interactions might be influenced by the mordanting method. More work is necessary to find out the reasons for the difference in color strength. Simultaneous mordanting has been adopted in all successive investigations.

18.8 Influence of pH on the dye uptake and color strength (K/S)

Trials have been carried out under various pH values using the simultaneous mordanting technique so as to study the influence of pH on the adsorption capacity of monascorubrin on silk fabric. The adsorption capacity reduced with pH increase ranging between 3 and 11. The possible mechanism of ionic interactions of the pigment anions with the protonated amino groups on the silk fiber suggests the possible influence of pH on adsorption of the monascorubrin. When the pH of the solution was under the isoelectric point of silk (approximately 5.0), the silk would have more positively charged sites through increased amino group protonation than at higher pH values [13]. There is a high negative potential for the monascorubrin molecule, i.e. at the hydroxyl and carbonyl group from an electrostatic potential map of monascorubrin. Owing to strong electrostatic attractions between the pigment and silk, the adsorption process is improved.

18.9 Rate of dyeing

The dyeing rate curves of monascorubrin pigment on silk fabric in the absence and presence of praseodymium chloride has been studied. The kinetic parameters of monascorubrin pigment have been determined. From the experimental kinetic data, the values of the time of half-dyeing ($t1/2$) have been estimated. The monascorubrin pigment dyeing rate alone has been lesser than for simultaneous mordanting with praseodymium chloride. In the case of the mordanted sample, the half-dyeing duration reduced and the equilibrium dye uptake increased drastically for the mordanted sample.

The results confirmed that the mordanted samples had higher dyeing capacity than the un-mordanted samples. The dyeing rate of monascorubrin pigment was highly dependent upon the affinity between monascorubrin pigment and silk fibers. Natural dyes often required metallic mordant to increase the affinity between the fiber and dyestuff. The mordant increased the net positive charge of the fiber. The ionic interaction between the anionic pigment and silk fiber correspondingly increased in the presence of mordant [13]. This electrostatic attraction plays an important role in enhancing the dyeability, which would increase the dyeing rate of monascorubrin pigment. The adsorption capacity with monascorubrin pigment itself was smaller than that observed for the terrae rare complex. The anion of the monascorubrin pigment had a complex characteristic. Terrae rare ions could function as bridging ligands, and enable to form coordination linkages between the pigment molecule and fiber. Hence there is increase in the rate of dyeing and equilibrium dye uptake.

18.10 Color-fastness

The washing-fastness, rubbing-fastness and light-fastness of the dyed samples with monascorubrin pigment with and without mordant have been determined. The findings reveal that the un-mordanted samples show low color-fastness. The low light-fastness is well known to be a problem with natural dyes [13]. The bad washing-fastness is due to the low affinity and fixation of monoascorubrin pigment onto silk fibers. The good fastness of the mordanted samples might be attributed to the covalent bond formation between the dye molecule, fiber and mordant. Terrae rare ions have been used to set pigments on fibers by forming an insoluble compound in order to improve the washing-fastness, rubbing fastness and light-fastness of the mordanted samples.

18.11 Build-up property

The build-up curves of monascorubrin pigment on silk fabric are explained. Evidently the equilibrium dye concentration on the fibers increased with the increase in the initial dye concentration. The equilibrium dye concentration in the fibers is a linear function of the dye concentration when the initial dye concentration is below 10% o.w.f [13]. The slope of the build-up curve becomes smaller when the initial dye concentration exceeded 10% o.w.f. The finding shows that monascorubrin pigment has excellent build-up for the silk fabric.

References

1. Dotan Y and Baslar S. Plants used as natural dye sources in Turkey. Econ Botany 2003; 57: 442–453.
2. Chairat M, Bremner JB and Chantrapromma K. Dyeing of cotton and silk yarn with the extracted dye from the fruit hulls of mangosteen, Garcinia mangostana Linn. Fibers Polymers 2007; 8: 613–619.

3. Mongkholrattanasit R and Wiener J. Dyeing and fastness properties of natural dyes extracted from eucalyptus leaves using padding techniques. Fibers Polymers 2010; 11: 346–350.
4. Unagul P and Wongsa P. Production of red pigments by the insect pathogenic fungus Cordyceps unilateralis BCC 1869. J Indust Microbiol Biotechnol 2005; 32: 135–140.
5. Kobrakov KI and Stankevich GS. The search for effective new dyes for chemical fibres. Fibre Chem 2005; 37: 84–88.
6. Lee D-K and Cho D-H. Fabrication of nontoxic natural dye from sappan wood. Korean J Chem Eng 2008; 25: 354–358.
7. Mati E and de Boer H. Contemporary knowledge of dye plant species and natural dye use in Kurdish autonomous region. Econ Botany 2010; 64: 137–148.
8. Mansour Heba F and Heffernan S. Environmental aspects on dyeing silk fabric with Sticta coronata lichen using ultrasonic energy and mild mordants. Clean Technol Environ Policy 2011; 13: 207–213.
9. Lee DK, Lee JH and Shin HY. Fabrication of nontoxic natural dye from sappan wood. Korean J Chem Eng 2008; 25: 354–358.
10. Lee YH and Kim HD. Dyeing properties and colour fastness of cotton and silk fabrics dyed with Cassia tora L. Fibers Polymers 2004; 5: 303–308.
11. Mohamed MS and Mohamad R. Enhancement of red pigment production by Monascus purpureus FTC 5391 through retrofitting of helical ribbon impeller in stirred tank fermenter. Food Bioprocess Technol 2009; 10: 214–219.
12. Babitha S, Julio C and Pandey A. Effect of light on growth, pigment production and culture morphology of Monascus purpureus in solid-state fermentation. World J Microbiol Biotechnol 2008; 24: 2671–2675.
13. Yanchun L, and Ganga Bai. Bioecological mordant dyeing of silk with monascorubrin, Textile Research Journal, (2011) 82(2)203.

19

Influence of laundering and water temperature on the properties of silk and silk-blended knits

Summary : Silk and silk-blended knits have been investigated and found to behave similarly to other fabrics as reported earlier in achieving dimensional stability by six laundering cycles. The temperature of water had a small influence on the behavior of the fabric over subsequent laundering cycles, but did not influence the 'stabilized' properties. It is considered important to manufacturers and consumers, since consumers may not always follow recommended laundering procedures. Also, most changes in fabric properties (with the exception possibly of dimensional changes) have been negligible and likely to be imperceptible to the consumer. But, some care needs to be exercised in maintaining silk knit garments, since silk fibers will degrade even after as few as six machine laundering cycles, as shown by the fibrillation of silk fibers through SEM photographs.

19.1 Introduction

The apparel market is influenced by innovation and provides consumers with an increasing array of products for selection. It is valid since fibers like "bamboo" (i.e. regenerated cellulose of bamboo origin) enter the market, and other established fibers (such as silk) have been used in non-traditional ways. There are constrains in use of typical fibers in apparel. Problems relating to concern about the physical properties of fabrics, and lack of knowledge and/or experience on how to clean and maintain garments, could pose problems in purchase. For example, silk has traditionally been used in formal wear (i.e. evening gowns) often made from delicate fabrics. The association between fragile, luxury fabrics and silk has contributed to claims that silk fabrics are both dimensionally unstable and difficult to care for (i.e. dryclean only). These issues hinder the adoption of silk fabrics for use in everyday applications such as next-to-skin knitwear. To encourage adoption of silk for such applications, consumers need to be informed about performance and care, and how to maintain these garments. Dimensional stability of fabric is important to both manufacturers and consumers, as unacceptable dimensional change is often cited as one of the primary reasons for consumer returns of textiles products [1]. While some dimensional change is needed for successful product manufacture, the inherent dimensional change in a fabric should be less than a pre-

DOI: 10.1201/9781003735427-19

described maximum (consistent between and within batches, and among the same colors of the same fabric), predictable and thus manageable. In knit fabrics, dimensional change is a normal part of the manufacturing process as yarns relax and loops reorient [2, 3]. Changes are exhibited as either shrinkage or expansion; however, the predominant source of dimensional change in knitted fabrics is relaxation of yarns. Owing to the extensible nature of knit fabrics, the limits of dimensional change are not considered as critical in knits as they are in with an increasing array of products to choose from. This is especially true as fibers such as "bamboo" (i.e. regenerated cellulose of bamboo origin) enter the market, and other established fibers (such as silk) are used in non-traditional ways. The use of atypical fibers in apparel can be problematic. Issues such as concern about the physical properties of fabrics, and lack of knowledge and/or experience on how to clean and maintain garments, may be barriers to purchase. For example, silk has traditionally been used in formal wear (i.e. evening gowns) often made from delicate fabrics. The association between fragile, luxury fabrics and silk has contributed to claims that silk fabrics are both dimensionally unstable and difficult to care for (i.e. dryclean only). These issues hinder the adoption of silk fabrics for use in everyday applications such as next-to-skin knitwear. To encourage adoption of silk for such applications, consumers need to be informed about performance and care, and how to maintain these garments. Dimensional stability of fabric is important to both manufacturers and consumers, as unacceptable dimensional change is often cited as one of the primary reasons for consumer returns of textiles products [1]. While some dimensional change is needed for successful product manufacture, the inherent dimensional change in a fabric should be less than a pre-described maximum (consistent between and within batches, and among the same colors of the same fabric), predictable and thus manageable. In knit fabrics, dimensional change is a normal part of the manufacturing process as yarns relax and loops reorient [2, 3]. Changes are exhibited as either shrinkage or expansion; however, the predominant source of dimensional change in knitted fabrics is relaxation of yarns. Owing to the extensible nature of knit fabrics, the limits of dimensional change are not considered as critical in knits as they are in woven fabrics. Limits of ±8% × 8% are cited as appropriate maximums for tubular knits (see e.g., [4]). Fabric characteristics, such as stitch length, machine stitches per centimeter, yarn twist, knitting tension, and laundering methods, may also affect stability [5]. Most dimensional change attributed to relaxation shrinkage occurs during the first laundering cycle, with fabrics progressively changing with each additional cycle until stability is achieved [6, 7]. If fabrics have not been sufficiently relaxed prior to garment manufacture, dimensional change will likely occur during laundering, resulting in differences between the size of the garment originally purchased and the stabilized garment. Therefore, critical to garment production is the process of stabilization (the number of wash cycles required to dimensionally stabilize a fabric) and the ability to predict the magnitude of any changes. A wide range of fabrics (varying in fiber types and structures) achieve dimensional stability after six consecutive wash cycles [6]. In contrast, silk knit fabrics may require a greater number of laundering cycles to achieve a fully relaxed state [8]. The temperature of the water used in laundering affects the efficiency of cleaning agents and is therefore a key variable in the cleaning process. However, water temperature does not necessarily have an effect on dimensional stability (e.g. cotton fabrics [9, 10]). In silk fabrics, examined at temperatures of 20°C, 35°C, and 50°C, most shrinkage was observed in fabrics laundered in water of 35°C. Such differences were attributed to temperature sensitivity of the silk fibers [3]. The silk yarns tend to develop hairiness after 10 washes (also caused by the temperature of the water used) [3]. The hairiness could indicate damage to the silk fibril due to mechanical action during laundering [11–14]. The aims of the research have been as follows:

1. The number of laundering cycles required for silk and silk blend knit fabrics to achieve stability; and
2. Laundering and wash water temperature on selected fabric properties during repeated laundering cycles. The current study is important because knitted fabrics made from specialty protein fibers such as silk are becoming increasingly popular for products intended for everyday consumer use.

Technical details

Silk and silk-blend knits of three commercial types (available as both black and non-dyed colorways) have been studied (Table 19.1). The physical properties evaluated were mass per unit area, and thickness [15, 16]. Dimensional stability was determined, percentage change was calculated, and stitches/cm measured [17, 18]. Specimens of 500 × 500 mm (n = 3 repetitions) were cut to ensure that none contained the same course or wale yarns [19]. The chosen fabric properties have been measured on "new" specimens (non-washed) and washed specimens (after each consecutive laundering cycle). Laundering cycles (i.e. wash, dry, press, condition, and assess) continued till fabrics attained dimensional stability. Fabrics were determined to be dimensionally stable when no changes of >1% occurred from one wash cycle to the next.

Table 19.1 Fabrication and technical information [29]

	Silk jersey blend	Silk jacquard blend	100% silk jersey
Fabric description Fiber content	72% spun silk, 15% cotton, 13% wool	80% spun silk, 20% wool	100% silk
Mass g/m^2	125	180	110
Structure	Single jersey	Jacquard	Single jersey
Yarn	120s	120s	120s
Colors	Non-dyed, Black, blue	Non-dyed, black	Non-dyed, black
Finishing			
Dyestuff	As per the European standard of environment protection		
Finishing	Anti-wrinkle, non-formaldehyde dye fixing agent, ammonium silicone softening agent, organo silicon softening agent		

New control fabric specimens were included in each wash cycle as a means of evaluating the consistency of the wash process when samples were matched in terms of wash temperature (i.e. among loads and over time). Specimens have been laundered as per ISO standards with 100% polyester fabric ballast making up a total load of 2 kg dry mass. Samples of each of the fabrics were subjected to wash as per ISO standards with the exception that wash and rinse water temperature were adjusted [20]. Specimens were subjected to one of three different washing cycle water temperatures (20°C, 35°C, or 50°C) and a standard rinse temperature of 20°C. In the absence of a standard specifying wash, water temperatures associated with descriptions of cold, warm and hot wash, wash cycle temperatures were determined. A standardized detergent has been used for all loads. After each wash cycle, specimens were dried flat under ambient conditions, pressed for 2 minutes using a domestic iron set to "high" (approximately 150°C), and conditioned for at least 24 hours in the standard atmosphere prior to measuring of physical properties [21]. New fabrics and those subjected to six wash cycles at 20°C, 35°C or 50°C were examined under microscope. In order to obtain a representative sample of images across the surface of the sample fabrics, five images were obtained at ×200 magnification randomly across the surface of each fabric. The characteristics of each fabric and treatment were then evaluated and described following review of the sample images by two judges working in a gray space under daylight conditions. ANOVA technique has been used for all laundering cycles. Means and standard deviations were calculated for all properties. Where differences among fabrics were found, Tukey's honestly significant difference (HSD) test was used to establish which means (calculated over all cycles) differed significantly at the $p \leq 0.05$ level [22]. Differences between new and stabilized fabrics have been calculated and then examined using univariate ANOVA [23].

Findings of the investigation

The thickness, mass, and number of stitches per centimeter of new non-laundered fabric and that subjected to six laundering cycles have been measured. The water temperature influences the dimensional change of the reference fabric (cotton duck) in warp and weft direction. But, the extent of shrinkage among the control fabrics subjected to the same wash water temperature did not vary over time implying no significant difference among washing cycles suggesting difference among silk fabrics is because of wash water temperature. The number of laundering cycles was the most important factor affecting fabric mass, thickness and the number of stitches per centimeter [29]. However, the type of fabric also had a large effect on a number of properties. Thickness differed depending on fabric type. The silk/cotton/wool jersey blend (SCW-J) and silk jersey (S-J) increased by a similar percentage, 2.3% and 2.1%, respectively. The silk/wool jacquard blend (SW-JQ) increased by 10% from new (non-washed) fabric to that of the dimensionally stable fabric (i.e. after six washes). Such differences in properties between fabric types were expected given the variations in fabric structures (single jersey versus jacquard). Change in mass over time also varied according to the type of fabric.

By the sixth laundering cycle, the mass of all fabrics had decreased from that when new by 3–13% per square meter (0.2–0.6% per sample). However, while the laundering process affected the mass of the fabric, the percentage change was sufficiently small to be considered negligible from the consumer and manufacturer's perspective. Of interest, however, is the reason for the changes in mass, and the pattern of change exhibited by the fabrics (Table 19.2). Possible explanations include changes in moisture retention, dimension and/or fiber loss due to abrasion/degradation during washing [13, 14]. When dimensional change was examined, differences in the way in which the fabrics behaved over time were apparent in both the course and wale directions. Dimensions of the fabric in the course direction decreased as the number of washes increased. For all fabrics, the greatest change occurred in the first wash cycle with the SW-JQ and the S-J exhibiting some recovery towards the original dimensions after subsequent washes. The SCW-J exhibited the greatest dimensional change (approx. −6.0%) and no observable recovery. In the wale direction, all three fabrics decreased in dimension, with the extent of change again varying among fabrics. The SCW-J had the smallest overall change in dimension in the wale direction (−0.5%), next was the S-J (−2.7%), followed by the SW-JQ (−4.5%). All fabrics behaved differently, most likely due to varying construction (the SW-JQ having a jacquard construction included comparatively longer loops in the wale direction, thus the greater degree of potential for shrinkage corresponded with tightening of these loops). Results from the present study support the contention of Gore et al. [6] that fabrics, including those made from the silk and silk blends that were examined here, are dimensionally stable after six laundering cycles. No evidence of greater instability in silk fabric, and thus a need for further wash cycles [3, 8], was identified. The effect of cleaning also varied according to the color of the fabric. The principal effect was that the non-dyed fabrics changed more (average over all washes) than the black fabrics when washed. Thickness (non-dyed 8.7%, black 1.1%), mass per unit area (non-dyed –0.4%, black 0.0%), dimension in both the wale and course directions (non-dyed –3.2%, black –2.0% wale; non-dyed –5.6%, black – 0.1% course direction); and stitches per centimeter (non-dyed 7.1%, black 2.2%) all differed depending on the color of the fabric. Although it is unlikely consumers would be concerned about such small changes in mass (<2 g/m^2) and thickness (<0.1 mm), inconsistent dimensional changes exhibited in garments constructed from different colors of the same fabric is of concern. In addition, dimensional change of the magnitude exhibited by some of the non-dyed fabric exceeds those considered the allowable maximum in industry (e.g. ±8% × 8% for tubular knits) [4]. Differences in dimensional stability between non-dyed and dyed fabric may reflect the stabilizing effect of dyeing occurring at the fabric stage.

If this is the case, the non-dyed fabric would benefit from an additional wet processing prior to use in apparel production. Of particular interest was the effect of wash water temperature on fabric properties. Consistent with findings reported elsewhere [9, 10], wash water temperature had only a very small effect on the pattern of relaxation and progressive shrinkage, and fabric mass, over the six laundering cycles, but did not affect the final dimensions ($F2,84 = 0.09$, NS wale direction; $F2,84 = 0.04$, NS course direction; $F2,84 = 0.06$, NS change in mass). Nor was there any evidence that the magnitude of dimensional change varied according to the wash temperature [3]. While the same wash temperature was used as in the study of Quaynor et al. [3], some notable differences in approach exist [3]. That is the washing machines used differed (Wascator versus domestic twin tub), as did the rinse temperatures (20°C versus not-specified); fabric type (commercially manufactured fabrics comprised of spun (staple-length) silk versus silk filaments on a hand knitting machine) and drying method (dried flat versus tumble drier) [3]. The effect of the tumble drier and the associated tumbling action on the dimensional stability and distortion of silk fabrics is not known; however, it does have a demonstrated effect on the stability of cotton fabrics [24]. Furthermore, tumble drying of commercially available silk garments is generally discouraged by manufacturers and retailers. Examination of the physical properties over time revealed some changes related to laundering water temperature. The pattern of percentage change in fabric thickness varied according to the wash water temperature. The warm (35°C) wash was associated with the greatest increase in mean fabric thickness (7.1%) with this increase in thickness being greatest for the non-dyed SW-JQ (19.0%). However, when thickness of the new non-washed fabrics was compared with that of the stabilized fabrics (i.e. after six washes) no significant difference was found between fabrics washed in hot, warm and cold wash water. Overall, water temperature did not significantly affect properties of the stabilized fabric, rather the pattern of change to stability varied among wash cycles.

Examination of the pattern of change raises questions as to whether water temperature was causing some effect at a fiber level, i.e. warm (35°C) wash water was associated with pattern aberration in both change in thickness and change in mass. Scanning electron microscopy (SEM) images of the fabrics and comparison of images of new fabric with those of fabric that had been washed six times revealed some evidence of longitudinal fibrillation and peeling of the silk fibers. Fibrillation and peeling were present in all fabrics, irrespective of whether new or subjected to repeated laundering. Fibrillation was observed on some fibers only, but was present in all fabrics (jacquard and jersey; new and washed), and all wash water temperatures. Most degradation appeared on surface fibers, located on the most exposed surfaces. Some differences among fabric types were identified.

The SW-JQ specimens exhibited the most extensive fibrillation possibly due to the more textured (raised) structure, providing a larger surface area and therefore greater potential for abrasion/degradation than the jersey fabrics. However, it is also possible that fabrics which are in fact more degraded may appear less damaged (i.e. exhibiting less fibrillation) due to fiber breakage and loss, which may also be indicated by the decrease in fabric mass. Quaynor et al. [8] commented on "hairy" yarns being identified on silk fabrics laundered in warm (35°C) water. In the SEM images examined here, fibrillation was clearly apparent at ×200 magnification [29]. Whether this "hairiness" was in fact fibrillation of individual silk fibers at a lower magnification is unknown. Such hairiness (if due to mechanical abrasion during processing) may have been more apparent than that occurring during processing and treatment of the fabrics examined during this study, which used commercially available fabrics which were dried flat between wash cycles. The tumbling action of a tumble dryer has been reported previously as having an important effect on the dimensional stability and distortion of weft knit fabrics [24]. Therefore, the method of washing and drying used may have contributed to differences in dimensional stability between silk and other fabrics claimed by Quaynor et al. [8].

It has been earlier reported that the tendency of manufactured cellulosic fibers to fibrillate was strongly related to fiber swelling properties, suggesting fibrillation is induced by mechanical abrasion in wet conditions [25, 26]. Silk-like cotton is a natural microfiber and is constructed from smaller fibril components which form the larger unit [12]. The fibroin core of the silk filament is a highly orientated and crystalline fibrous structure which, while water insoluble, is very absorbent and readily wetted out by water [27]. Associated fiber swelling and fiber diameter increases of 16–18% result and when the fibers are wet the fibrils are easily separated by physical force such as that exerted during washing [11, 13]. The fibroin core of silk is protected to some extent during processing by the enveloping sericin sheath formed around the fibroin core during extrusion of the fiber [12]. During processing, removal of sericin occurs during the degumming process (after weaving or knitting) at which time the fibroin core is exposed to mechanical action during wet processing and dyeing [12].

Mechanical damage of the silk fiber surface is related to a dull appearance and impairment of final quality [12]. Whether the abrasion apparent on the fabric surfaces examined was affected by wash water temperature requires further scrutiny, as the effect of water temperature on durability and appearance of silk knit fabrics over time (greater than six wash cycles) is not known. In most countries, garments are required to carry care labels which indicate care procedures including wash methods and temperature [28]. Irrespective of whether labels on silk garments recommend hand or machine washing, it is the consumer who ultimately controls wash temperature. Variations between hot water heater settings and the temperature of incoming cold tap water mean that water temperatures may vary widely between domestic washing machines [29]. Also, washing machine cycles and care label do not relate to a specific temperature indicating only "cool/cold", "warm", or "hot" with the interpretation of such descriptions differing among consumers and countries, etc. When silk fibers are affected by water temperature, the use of terms like "cool", "warm", and "hot" is likely to contribute to fiber and thus garment degradation and eventually lead to displeasure of the customer.

References

1. Powderly, D. D. Dimensional changes in fabric and apparel related to home laundering practices, *Textile Chem. Color.*,**10**(8), 27–31 (1978).
2. Kadolph, S. J., Langford, A. L., Hollen, N., and Saddler, J., *Textiles*, 7th edn, Macmillan Publishing Co, New York, 1993.
3. Quaynor, L., Takahashi, M., and Nakajima, M. Effects of laundering on the surface properties and dimensional stability of plain knitted fabrics, *Textile Res. J.*, 70(1), 28–35 (2000).

4. Mehta, P.V., and Bhardwaj, S.K., *Managing Quality in the Apparel Industry*, New Age International Publishers, Bangalore, 1998.
5. Suh, M.W. A study of the shrinkage of plain knitted cotton fabric, based on the structural changes of the loop geometry due to yarn swelling and deswelling, *Textile Res. J.*, **37**(5), 417–431 (1967).
6. Gore, S.E., Laing, R.M., Wilson, C.A., Carr, D.J., and Niven, B.E. Standardizing a pre- treatment cleaning procedure and effects of application on apparel fabrics, *Textile Res. J.*, **76**(6), 455–464 (2006).
7. Higgins, L., Anand, S.C., Holmes, D.A., Hall, M.E., and Underly, K. Effects of various home laundering practices on the dimensional stability, wrinkling, and other properties of plain woven cotton fabrics. Part 1: Experimental overview, reproducibility of results, and effect of detergent, *Textile Res. J.*, **73**(4), 357–366 (2003).
8. Quaynor, L., Takahashi, M., and Nakajima, M. Dimensional changes in knitted silk and cotton fabrics with laundering, *Textile Res. J.*, **69**(4), 285–291 (1999).
9. Morris, M.A. Laundering cotton fabric: Part II: Effect of detergent type and water temperature on appearance, hand, strength and cost, *Textile Res. J.*, **40**(7), 644–649 (1970), http://trj.sagepub.com/cgi/content/abstract/40/7/644.
10. Powderly, D.D. Effect of water temperature of dimensional change in laundering, *Textile Chem. Color.*, **12**(10), 260–261 (1980).
11. Goynes, W.R., and Rollins, M.L. A Scanning Electron-Microscope Study of Washer-Dryer Abrasion in Cotton Fibers, *Textile Res. J.*, **41**(3), 226–231 (1971) http://trj.sagepub.com/cgi/content/abstract/41/3/226.
12. Freddi, G., Mossotti, F., and Innocenti, R. Degumming of silk fabric with several proteases, *J. Biotechnol.*, **106**, 101–112 (2003).
13. Handu, J.L., Sreenivas, K., and Ranganathan, S.R. Chemical and Mechanical Damage in Service Wear of Cotton Apparel Fabrics, *Textile Res. J.*, **37**(11), 997–999 (1967).
14. Rollins ML, Degruy IV, Hensarling TP, and Carra JH. Abrasion Phenomena in Durable-Press Cotton Fabrics: A Microscopical View. *Textile Res. J.*, **40**(10), 903–916 (1970), http://trj.sagepub.com/cgi/content/abstract/40/10/903.
15. British Standards Institution. BS EN 12127:1998 Determination washing and drying. International Organization for Standardization, Geneva, 2007.
18. British Standards Institution. BS 5441: British Standard Methods of test for Knitted Fabrics. British Standards Institution, London, 1988.
19. International Organization for Standardization. ISO 3759: 2007 (E) Textiles—Preparation, marking and measuring of fabric specimens and garments in test for determination of dimensional change. International Organization for Standardization, Geneva, 2007.
20. International Organization for Standardization. EN ISO 6330:2000 Textiles—Domestic washing and drying procedures for textile testing. International Organization for Standardization, Geneva, 2000.
21. International Organization for Standardization. EN ISO 139:2005 Textiles—Standard atmospheres for conditioning and testing. International Organization for Standardization, Geneva, 2005.
22. Zar, J.H. Biostatistical Analysis, 2nd Edn, New Jersery, Prentice-Hall Inc, 1984.

23. SPSS Inc. SPSS version 16.0.2 for Windows. SPSS Inc., Chicago, IL, 2008.
24. Higgins, L., Anand, S.C., Hall, M.E., and Holmes, D.A. Factors during tumble drying that influence dimensional stability and distortion of cotton knitted fabrics, *Int. J. Cloth. Sci. Technol.*, **15**(2), 126–139 (2003).
25. Zhang, W., Okubayashi, S., and Bechtold, T. Modification of fibrillation by textile chemical processing, *Lenzinger Berichte*, **82**, 58–63 (2003).
26. Okubayashi, S., and Bechtold, T. A pilling mechanism of man-made cellulosic fabrics—effects of fibrillation, *Textile Res. J.*, **75**(4), 288–292 (2005).
27. Mishra, S.P. *A Textbook of Fibre Science and Technology*, New Age International Publishers, New Delhi, 2000.
28. Rudkin, A. 'Care labelling', http://www.intertek-labtest.com/brochures/CareLabelling (2006, accessed 13 May 2008).
29. Rebecca R. Van Amber, Brian E. Niven and Cheryl A. Wilson. Effects of Laundering and Water Temperature on the Properties of Silk and Silk-blend Knitted Fabrics, Textile Research Journal, Vol. 80(15): 1557.

20

Physical and thermal properties of grafted silk fabrics

Summary : The graft copolymerization method has been used to modify silk fabrics with phosmers to achieve better flame retardancy. Techniques like SEM, FTIR, DSC, TG, TMA and tensile measurements have been used to measure the surface morphology, molecular structure, thermal behavior and mechanical properties of fire-retardant silk fabrics. In the case of phosmer CL, the weight gain of the silk fabric grafted exhibited higher graft polymer efficiency than the phosmer M. The grafting of silk fabrics with phosmers grafting agents imparted flame retardation potentiality and permitted preparation of flame-retardant fibers. The flame retardancy of silk grafted with phosmer CL was higher and there was a longer time to ignition, 120 seconds, whereas it took 5 seconds for the silk grafted with phosmer M to ignite. SEM images showed that the surface of the grafted silk fabrics was smooth and flat as that of control silk fabrics. The FTIR result indicated that phosphate groups were introduced within the matrix of silk fibers. DSC results suggested that the position of the major endothermic peak of the silk fabric grafted with phosmer CL and M shifted toward higher temperatures, 327°C and 338°C, respectively, suggesting the high thermal stability of grafted silk fibers. The flame retardation of silk fiber after grafting should be partly related to the shifting of the thermal decomposition of silk fibroin. The fabric tensile strength loss is negligible in the grafted silk. The grafted silk structure is unaltered with a fireproof protective layer at the surface. There appears to be no necessity to attain high levels of weight gain for the fabrics to show a high degree of flame retardation. Thus, fabrics having good performance as flame-retardant silk fibers can be prepared by controlling the weight increase between 5% and 25%, which is the safest for maintaining the intrinsic silk fabric properties constant. Flame-retardant silk fabrics find uses in the production of specific clothing articles and also in technical textiles.

20.1 Introduction

An important functional property of clothing is the extent of the fire retardation of textile material. This proves that the flame-retardant clothing is suitable in uniforms for the firemen and steel workers. The method of fireproofing becomes crucial considering the production of fire-retardant textile fabrics. But, it is impractical to produce perfectly inflammable fabrics comprising of organic and natural fibers on the one hand, and on the other hand glass fiber and asbestos show perfect flame retardation. Halogen compounds are generally used as additive-type flame retardants. Cotton treated with alginic acid derivatives of alkali metal renders it fire

DOI: 10.1201/9781003735427-20

retardation [1]. The carboxymethylation of cotton treated with acrylamide in alkaline state is one of the techniques of introducing carboxy groups into the cotton fibers, and metal-cation-exchanged carboxyethylation is effective on cotton [2, 3]. The research on thermal properties of Nomex/ viscose fiber blended fabric has been conducted. It is revealed that most of the blended materials demonstrated good fire-retardant behaviors, the blending ratio of 80/20 (Nomex/ Lenzing Viscose FR) showed excellent thermal properties [4]. There have been studies on the utilization of hydroxyl-functional organophosphorus oligomers and the binders for flame-retardant finishing of nylon and cotton fabric [5, 6]. However, major problems encountered with these additives concern the generation of toxic and corrosive fumes during combustion. To circumvent these problems, non-halogenated flame retardants would be potentially important chemical agents of choice. Molecules containing phosphorous groups belong to these fire-retardant agents allowing improved material thermal and fire properties [7]. Phosphorus, silicon and nitrogen compounds are the agents mostly used for replacing halogenated compounds in flame-retarding resins, especially epoxies [8]. The halogen-free products are attractive and welcome worldwide [9]. It is of interest to note that the chemical treatment and flame-retardant additives produce an after effect on the textile materials, such as decrease the mechanical properties [10]. The durability of flame retardation should be considered for practical usages. The de novo techniques for producing flame retardation are highly desired, with excellent durability of flame retardation and without reducing the original textile functional performances. Silk fabric is a luxurious fashionable apparel used for interior decorations, such as upholsteries, curtains, and beddings for its soft handle, gloss, fineness, higher strength, high wearing comfort, and other desirable aesthetic appearance. Therefore, flame retardancy is an important and desirable property that consumers can expect for silk textiles. It is well known that the properties of silk such as dimensional stability, wash and wear, wrinkle recovery are inferior. Grafting is considered as an effective technique to improve the performance of silk fibers and fabrics. These techniques are promising to be a powerful method for producing substantial modification of fiber properties as well as improving their textile performances [11, 12]. The grafting monomers such as methacrylamide (MAA), 2-hydroxyethyl methacrylate (HEMA), and grafting binary system using MAA and HEMA had been successfully applied on the industrial scale for silk processing [13–16]. The physical and chemical properties of grafted silk fibers are widely examined as a function of fiber weight gain. But, the flame-retardant finishing technology available for silk is limited. Little work has been reported during recent years, on the flame retardant modification of silk fabrics using graft polymerization techniques.

A vinyl phosphorous based monomer, dimethyl 2-(methacryloyloxyethyl) phosphate (DMMEP), diethyl 2-(methacryloyloxyethyl) phosphate (DEMEP) and organophosphorus oligomer were applied onto silk for fire-retardant modifications and achieved good silk flame retardant performance [17–19]. Argon plasma jet was used to graft a non-durable phosphorous-based flame retardant agent onto silk [20]. The treated fabric showed a higher level of flame retardancy as evaluated by using the 45° flammability test. But, elaborate studies are so restricted and no work has been reported relating to the application of phosmer M and phosmer CL onto silk fabrics. Two new flame retardant agents have been used with phosphoric groups: phosmer M and phosmer CL and applied on silk fabric so as to achieve high performance fire retardant silk fabrics. The thermal characteristics, thermo mechanical properties, tensile strength, surface morphology and molecular structure of flame retardant silk fabrics have been analyzed.

Technical details

Degummed habutae silk fabric (plain weave) has been used as the substrate for the material. Silk fabrics have been grafted with grafting monomers with phosphoric groups, i.e. phosmer M and phosmer CL. The proper chemical names for the phosmer M and CL are acid phosphoxyethylmethacrylate, and 3-chloro 2-acid phosphoxypropylmethacrylate, respectively. The grafting reaction of phosmer on silk fabric has been done as per earlier reported procedure [21]. Silk fabrics have been immersed in a grafting solution containing 1.0 g/L non-ionic emulsifying agent (polyoxyethylenealkylphenylether), 3% owf (on the weight of fibers) ammonium persulfate, and 120% owf phosmer. The pH of the grafting system was adjusted to 3.0 by adding 0.5 g/L formic acid solution. A material to liquor ratio of 1:30 was maintained for the course of the reaction. The grafting system was heated from room temperature to 80°C over 45 min and was maintained at this temperature for 30 min. At the end of the reaction, the treated silk was soaked in a solution containing 1.0 g/L nonionic detergent at 80°C for 30 min, and rinsed with water thoroughly. Un-reacted phosmer contained in the silk fabrics was removed by treatment with acetone at 55°C for 1 h and then with 1.0 g/L nonionic emulsifying agent solution at 80°C for 30 min. The specimens were then washed three times with hot water and air dried at 100–105°C for 2 h. The add-on percentage (Wg), by graft copolymerization of silk with phosmers, was calculated from the increase in the weight of the silk using the following equation:

$$W_g = \frac{W_2 - W_1}{W_1} \times 100$$

where W_1 and W_2 are the weights of the dried specimens before and after grafting, respectively. A correction was made for the fiber weight loss in the reaction system that occurred during the treatment by a comparison with a blank sample treated without monomer. The morphologies of the fibers were examined. Fourier transform infrared (FT-IR) spectroscopy was measured with a Shimadzu FT-IR-8400S infrared spectrometer by the attenuated total reflection method in the region of 4000–400 cm^{-1} at room temperature. Thermogravimetric (TG) analysis was carried out under nitrogen gas at a heating rate of 10°C/min. Differential scanning calorimetry (DSC) measurement was performed at a heating rate of 10°C/min under nitrogen gas atmosphere. An apparatus for thermo mechanical analysis (TMA) was used to detect the thermal expansion and contraction properties in the course of the heating process. The temperature range studied was from 25°C to 350°C. The heating rate was 10°C/min, and dry nitrogen gas provided the inert atmosphere. The TMA full scale was 500 mm. The initial load applied to the sample was 1 g. The tensile properties have been determined by using a standard technique at 22°C and 65% RH at a gauge length of 15 mm and a strain rate of 10% per min.

Flame-retardant behaviors of the grafted silk fabrics have been tested. As per the technique, the silk fabric (20 cm × 20 cm) was attached on the inclined sample stage of the burner at 45° angle. The flame has been impinged till the fabric started to ignite by the gas burner flame.

Grafting of silk fibers with phosmers

For controlling the degree of fiber grafting, the amount and type of monomer are considered crucial. The performance of phosmer CL and M have been assessed by calculating the weight gain% of silk fabrics grafted with both monomers at various durations and concentration of grafting. The fiber weight gain of silk fabrics graft copolymerized with phosmer M and CL as a function of grafting time is explained. Initially, the weight gain for the silk fabrics grafted with phosmers rapidly increases with increasing grafting time. After grafting for 40 min, weight gain grafted with phosmer M almost saturated or increased slowly, whereas weight gain grafted with phosmer CL further increased, reaching 13.9wt% and 21.1wt% after 90 min for phosmer M and CL, respectively. The influence of the monomer concentration on the weight gain of silk fabrics grafted with phosmer CL and M is defined. Initially, the weight gain of the silk fabric grafted with phosmer CL and M increases almost linearly with increasing the monomer concentration and reached at about 10wt% at the concentration of 50% owf.

Then, the fiber weight gain for the silk fabrics grafted with phosmer CL increased up to 21wt% at 120% owf phosmer concentration, whereas a further increase in phosmer M concentration had only a slight positive effect on the weight gain (&13.5%). These findings indicate that the phosmer CL showed higher graft polymer efficiency than the phosmer M. It could be due to the synergistic effect of chlorine group in the phosmer CL.

Structural characteristics

In an attempt to examine whether the grafting with different kinds of phosmers induced some structural changes, we investigated the changes in the secondary structure of silk grafted with phosmers CL and M.

The control silk fiber is characterized by the major absorption bands at 1620 cm^{-1} (amide I) and 1512 cm^{-1} (amide II), attributed to the b-sheet structure [22, 23]. It should be noted that the major absorption in the amide I and amide II band of grafted silk fibers with phosmer M and CL is the same as that of the control silk fiber. However, after grafting with phosmer M and CL, new absorption peaks appear at 1737 cm^{-1} and are attributed to the carbonyl stretching vibration of the flame retardant; at 1250 cm^{-1}, assigned to the P=O stretching vibration; at 1151 cm^{-1}, attributed to the P–O–C stretching vibration; and at 993 cm^{-1}, assigned to the P–O stretching vibration [18, 24]. The finding shows that phosphate groups have been introduced within the matrix of silk fibers. The free radical initiation technique is used to carry out graft copolymerization of vinyl monomers onto silk fibers. Free radical sites are formed on the backbone of the fibroin molecule at the primary stage. Vinyl monomer, such as phosmer M and CL, immerse in the matrix of silk fiber together with ammonium persulfate (APS), as the grafting initiator, graft copolymerize inside of silk fiber.

Surface morphology

Figure 20.1 shows SEM observation of the surface morphology of the silk fabrics grafted with phosmers M and CL. The degummed silk fibers have a very smooth surface, exhibiting only very fine longitudinal striation due to the fibrillar structure of the fiber [25]. The grafted silk fabrics have smooth and flat surface similar to that of control silk fabrics. No substance has been attached on the silk fabric surface after grafting with phosmer M and CL. These imply that there was no outstanding crack and that the fibers possess a dense inner structure after grafting with phosmers.

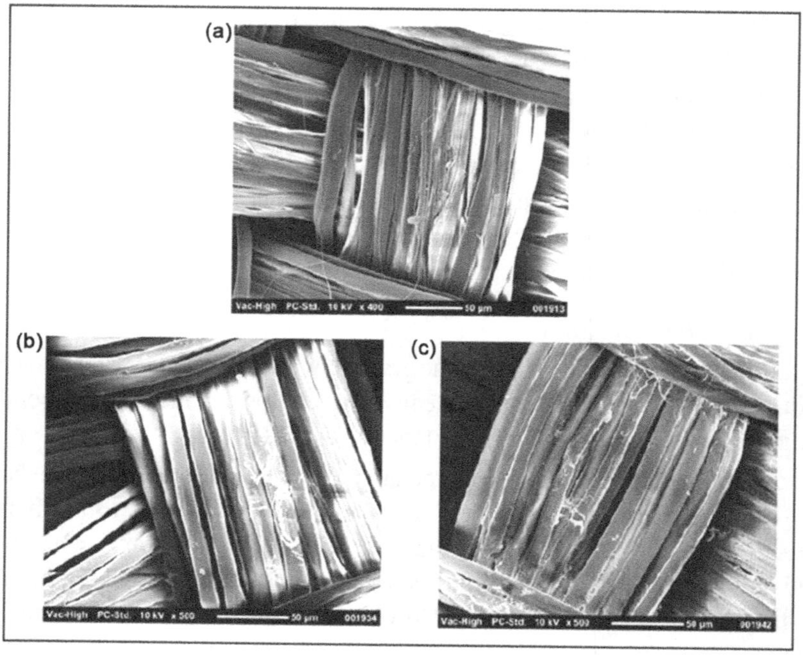

Figure 20.1 SEM photographs of silk fibers grafted with phosmers M and CL [28] (a) Control silk (b) Silk grafted with phosmer M, and (c) Silk grafted with phosmer CL

Flame retardant performance

The results of flammability test of grafted silk fabrics with phosmer M and CL have been studied. The ignition duration has been 2 and 5 seconds respectively in the case of the control and silk fabric grafted with phosmer M. But the ignition duration in the case of the silk fabric grafted with phosmer CL was significantly increased and consumed 2 minutes to burn. It is revealed that the after flame time for the control and the silk fabric grafted with phosmer M were about 2.9 and 2.1 second, respectively, and that for the silk fabric grafted with phosmer CL was 0, suggesting that the flame of the silk grafted with phosmer CL extinguished after the gas burner flame was removed. Also, the flame of the silk fabrics grafted with phosmer CL was self-distinguishing after removing the gas burner flame. The areas of the charred control silk fabric has been found to be 40.6 cm^2, and that for silk fabric grafted with phosmer M and CL has been found to be 28.2 cm^2 and 37.4 cm^2, respectively.

Thermal behavior

DSC measurement has been used to evaluate the thermal behavior of silk fibers grafted with different kinds of phosmer. The DSC thermograms of control and silk fibers grafted with phosmer M and CL are explained. A single predominant endothermic transition at 316°C characterize the DSC curve of control silk fiber, and is due to the thermal decomposition of silk fibroin with oriented b0 configuration [26]. The position of the major endothermic peak of the silk fabric grafted with phosmer CL and M shifted toward higher temperatures, 327°C and 338°C, respectively, suggesting the high thermal stability of grafted silk fibers. The most interesting feature is that the extent of the shifting toward higher temperature for the silk fiber grafted with phosmer M is larger than that of silk fiber grafted with phosmer CL. This result suggested the increased thermal molecular mobility of the silk fibroin molecules, as well as the phosmer M polymer grafted in the matrix of silk fiber, which is because of the less bulky side chains of the phosmer CL. These results could be linked to the increased flame retardation of the silk fibers. To elucidate the thermal characteristics of silk fibers grafted with phosmers, TG measurement was performed. A TG analysis was very useful to determine quantitatively the degradation behavior and the composition of the fiber.

TG and DTG (differential thermogravimetric) curves of silk fibers are grafted with phosmer M and CL with weight gain of 19.8 and 19.7, respectively. For the control silk fiber, the initial weight loss below 100°C was attributed to the evaporation of water, and was followed by nearly constant weight from 90°C to 250°C.

A second large weight loss took place in the temperature range from 260°C to 368°C. This was associated with the degradation of side chain groups of amino acid residues and the cleavage of peptide bonds [27]. TG results in the first step for the silk fabrics grafted with phosmer M and CL are similar and show that these samples are more stable than the control sample up to 259°C. Above 259°C, the TG curves exhibit one-step degradation. Between 259°C and 368°C, the thermal stability of silk fibers slightly increased after grafting. According to the DTG curves of grafted silk fibers, it is evident that the peak position corresponding to the thermal decomposition of silk fiber, shifted to the high temperature, from 323°C to 340°C after graft copolymerization. Above 368°C, all silk fibers degrade and a lower residual char is obtained for the grafted silk fibers. The quantity of residual char of the sample weight above 368°C is given as:

Control < silk fiber grafted with phosmer CL < silk fiber grafted with phosmer M.

The results indicate that the thermal decomposition of the graft polymer of phosmer M and CL onto the silk fiber matrix occurred in the narrow temperature zone.

Thermomechanical properties

The dimensional stability of silk fibers is considered an important functional property during the heat treatment. The thermal expansion and contraction properties have been measured by thermomechanical analysis in order to provide further information on the thermal behaviors of the silk fibers. The TMA curves of silk fibers grafted with phosmer Mand CL are explained. The control silk fiber showed a large contraction in the wide temperature range between room temperature and 200°C, following water evaporation [28]. At a temperature of about 200°C, the control silk sample exhibits rapid contraction, and then contracted more rapidly subsequently. The sample size of silk fiber grafted with phosmer M and CL stayed unchanged in the temperature range between room temperature and 300°C, when the sample's length contracted rapidly. The results exhibit that the size stability of the silk fabrics increased significantly after graft copolymerization.

Tensile properties

The tensile properties are considered crucial in determining the performance of fibers for suitable applications. The tensile properties of the silk fibers grafted with phosmers CL and M is important, since a grafting technique without tensile property modification would be an ideal method for using on textile materials. The tensile strength of silk fibers are grafted with various phosmers, CL and M. The tensile strength of silk fibers grafted with phosmer CL and M was not remarkably changed, whereas strain percentage was slightly decreased [28]. The stress value of control silk is 312, while for silk grafted with phosmers M, it is 276, and for phosmor CL, it is 273 MPa. It is evident that the grafting of silk fiber with phosmer CL and M does not drastically affect the sample tensile properties of the original sample. The existing tensile properties of silk are not destroyed by grafting.

References

1. Madeline RG and Pathus Labour JE. Fire proofing of textile and paper. US Patent 2,907,683 1959, 1959.
2. Reinhardt RM and Bruno JS. Carboxyethylation of cotton by treatment with acrylamide. J Appl Polym Sci 1966; 10: 387–397.
3. Mehta RD. Flammability of metal-cation-exchanged carboxyethylated cottons. Textile Res J 1974; 44: 825–826.
4. Huang G. Research on thermal properties of Nomex/Viscose FR fiber blended fabric. Materials and Design 2009; 30: 4324–4327.
5. Yang H and Yang CQ. Durable flame retardant finishing of the nylon/cotton blend fabric using a hydroxyl-functional organophosphorus oligomer. Polym Degrad Stab 2005; 88(3): 363–370.
6. Wu WD and Yang CQ. Comparison of different reactive organophosphorus flame retardant agents for cotton. Part I – The bonding of the flame retardant agents to cotton. Polym Degrad Stab 2006; 91(11): 2541–2548.

7. Lin JF, Ho CF and Huang SK. Thermal characterization of the phosphorus-containing sulfone-modified epoxy resins by thermogravimetric analysis and direct pyrolysis-GC/MS measurement on the thermally degradative volatiles. Polym Degrad Stab 2000; 67: 137–147.

8. Bergendhl CG. Alternative to halogenated flame retardants in electronic and electrical products. IVF research publication 99824. Moldal. Sweden: The Swedish Institute of Production Engineering Research, 1999.

9. Wu CS, Liu YL, Chiu YC and Chiu YS. Thermal stability of epoxy resins containing flame retardant components: an evaluation with thermogravimetric analysis. Polym Degrad Stab 2002; 78(1): 41–48.

10. Dolui SK. Unusual effect of filler ($CaCO_3$) on thermal degradation of polyurethane. J Appl Polym Sci 1994; 53: 463–465.

11. Lenka S. Grafting of vinyl monomers onto silk fibers. J Macromol Sci Part C: Polym Rev 1982; 22(2): 303–342.

12. Mohanty N, Torasia SN and Mohanty MC. Modification of tussah silk by methyl methacrylate. Part II. J Macromol Sci Part A: Pure Appl Chem 1983; 20(3): 409–419.

13. Kameda T and Tsukada M. Structure and thermal analyses of MAA-grafted silk fiber using DSC and solid-state NMR. Macromol Mater Eng 2006; 291: 877–882.

14. Tsukada M, Freddi G, Monti P, Bertoluzza A and Shiozaki H. Physical properties of 2-Hydroxyethyl methacrylate-grafted silk fibers. J Appl Polym Sci 1993; 49: 1835–1844.

15. Freddi G, Massafra MR, Beretta S, Shibata S, Gotoh Y, Yasui H and Tsukada M. Structure and properties of Bombyx mori silk fibers grafted with MAA and HEMA. J Appl Polym Sci 1996; 60: 1867–1876.

16. Tsukada M, Freddi G, Rosaria M, and Bretta S. Structure and properties of tussah silk fibers graft-copolymerized with MAA and HEMA. J Appl Polym Sci 1998; 67: 1393–1403.

17. Guan J and Chen G. Graft copolymerization of silk fabric with an organophosphorus flame retardant. Fire Mater 2010; 34: 261–270.

18. Guan J and Chen G. Flame resistant modification of silk fabric with vinyl phosphate. Fibers and Polymers 2008; 9(4): 438–443.

19. Guan J, Yang CQ and Chen G. Formaldehyde-free flame retardant finishing of silk using a hydroxyl-functional organophosphorus oligomer. Polym Degrad Stab 2009; 94(3): 450–455.

20. Chaiwong C, Tunma S, Sangprasert W, Nimmanpipug P, and Boonyawan D. Graft polymerization of flame-retardant compound onto silk via plasma jet. Surf Coat Technol 2010; 204: 2991–2995.

21. Tsukada M, Arai T, Winker S, Freddi G and Ishikawa H. Physical properties of silk fibers grafted with vinyltrimethoxysilane. J Appl Polym Sci 2001; 79: 1764–1770.

22. Shao J, Zheng J, Liu J and Carr CM. Fourier transform Raman and Fourier transform infrared spectroscopy studies of silk fibroin. J Appl Polym Sci 2005; 96(6): 1999–2004.

23. Bhat NV and Ahirrao SM. Investigation of the structure of silk film regenerated with lithium thiocyanate solution. J Polym Sci Part A: Polym Chem 1983; 21(5): 1273–1280.

24. Tsafack MJ and Levalois-Gru¨ tzmacher J. Plasma induced graft-polymerization of flame retardant monomers onto PAN fabrics. Surf Coat Tech 2006; 200(11): 3503–3510.

25. Arai T, Freddi G, Innocenti R and Tsukada M. Biodegradation of Bombyx mori silk fibroin fibers and films. J Appl Polym Sci 2004; 91: 2383–2390.

26. Ishikawa H, Tsukada M, Toizume I, Konda A and Hirabayashi K. DSC thermograms of silk fibroin. Sen-I Gakkaishi 1972; 28(4–5): 91–98.

27. Freddi G, Tsukada M and Beretta S. Structure and physical properties of silk fibroin/polyacrylamide blend films. J Appl Polym Sci 1999; 71: 1563–1571.

28. Masuhiro Tsukada, Md. Majibur Rahman Khan, Tomoko Tanaka and Hideaki Morikawa. Thermal characteristics and physical properties of silk fabrics grafted with phosphorous flame retardant agents, Textile Research Journal 81(15), 1541–1548.

21

Anti-wrinkle properties of silk fabrics dyed with reactive and crosslinking dyes

Summary : Studies on silk fabrics have revealed that C.I. Reactive Red 2, C.I. Reactive Red 195 and C.I. Reactive Red 141, and crosslinking dyes PAD1, PAD2 and PAD3 showed anti-wrinkle effects. Of these, bifunctional reactive dyes (C.I. Reactive Red 195 and C.I. Reactive Red 141) exhibited better anti-wrinkle properties than dichloro-s-triazine one (C.I. Reactive Red 2) since more cross linking was formed within silk fiber molecules. The anti-wrinkle properties of crosslinking dyes are superior in comparison with those of bifunctional reactive dyes owing to mainly the formation of a greater number of crosslinking bondings. Moreover, with the increase of the concentrations of both the cross linking dyes and the crosslinker, the wrinkle resistance of dyed silk fabrics was further improved. When the concentration of the cross linking dyes was 0.67 g/L and that of the crosslinker was 40 g/L, or when the dye applied was 4.0 g/L and the crosslinker was 20 g/L, the WRA and IRW of the dyed silk fabrics were comparable to those obtained from finishing with EPSIB and glyoxal. Moreover, the handle and tensile strength of silk fabrics continue to be satisfactory in these states. Hence, it could be made possible to perform simultaneous dyeing and finishing on silk fabrics by the application of the crosslinking dyeing process.

21.1 Introduction

Silk is an important natural fiber having the advantages of wearing comfort, soft handle, good air permeability and elegant appearance. But, silk fabric exhibit a demerit of crease when in storage and in wearing, which influences its wearability. Anti-wrinkle finishing agents are normally applied in after-treatment so as to overcome the problem. In the initial stage, trimethylol melamine has been used, though it is released as formaldehyde. Finishing agents like 1,2,3,4-butaneteracarboxylic acid (BTCA)1 have been evolved thereafter [5]. But, BTCA proves to be non-economical and needs high temperature during treatment. After BTCA treatment, the handle, draping property and the strength of the finished silk fabric get deteriorated significantly. For imparting anti-wrinkle finishes, epoxide EPSIB and glyoxal have also been used. The expoxide EPSIB and glyoxal suffer demerits given as below:

DOI: 10.1201/9781003735427-21

(a) A large quantity of agents is needed in application (EPSIB: 100 g/L and glyoxal: 96 g/L) [6, 7].

(b) The glyoxal usually harms the whiteness and strength of silk fibers, thus auxiliaries should be added to promote the reaction with the fibers.

There are two theories for the mechanism of anti-wrinkle finishing of silk fabrics. One is the deposition theory: it is proposed that small particles of finishing agents disperse in the amorphous region of the fibers where they deposit after resinification and form H-bonds with the fiber molecules, which restricts the movement of the molecular chains and decreases the deformation of the fabrics. The other is the crosslinking theory: it is proposed that finishing agents form covalent bonds with fiber molecules, and connect the adjacent molecular chains within the fibers, which enhances the recovery ability of the deformed fibers and results in increased wrinkle resistance. It has been found that most of the finishing processes for the anti-wrinkle agents can be explained by the crosslinking theory. In general, dyeing and the anti-wrinkle finishing of fabrics are separate processes. But, recent investigations have revealed that some dyes, like reactive dyes and crosslinking dyes, can form covalent bonds with silk fibers and enhance the anti-wrinkle property of the dyed fibers. The anti-wrinkle performances of the reactive dyes and crosslinking dyes with different molecular structures on silk fabric have been studied so as to possibly achieve simultaneous dyeing and finishing. The crosslinking dyes and cross linker used are economical and required no independent anti-wrinkle finishing process.

21.2 Technical details

Silk fabrics that are bleached, desized and mercerized 100% (plain woven) have been used. The dyestuffs used are C.I. Reactive Red 3, C.I. Reactive Red 2, C.I. Reactive Red 195 and C.I. Reactive Red 141. Polyethylene polyamine crosslinking dyes (PAD1, PAD2 and PAD3) have been prepared according to reported techniques [8, 9]. Crosslinker 2-chloro-4,6-di(aminobenzene-4′-β-sulfatoethyl-sulfone)-1,3,5-*s*-triazine (XLC) has been produced as per the technique elaborated by some research workers [10,11]. All the other chemicals conforming to analytical standards have been used [12]. The structures of the reactive dyes, crosslinking dyes and crosslinker XLC are given in Figures 21.1–21.3, respectively.

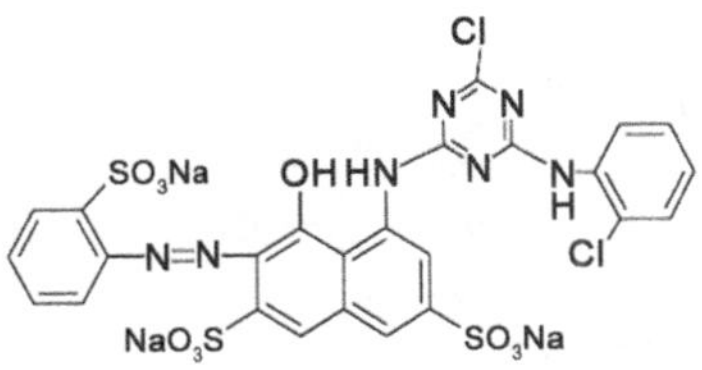

C.I. Reactive Red 3

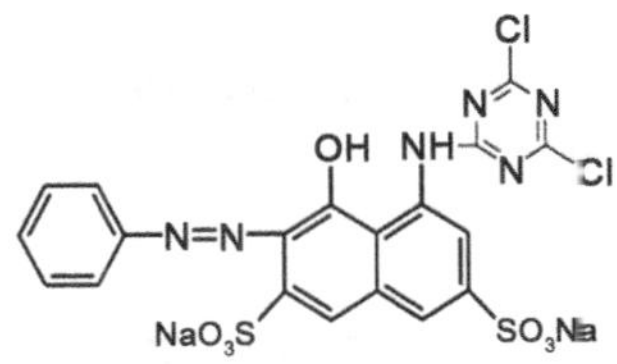

C.I. Reactive Red 2

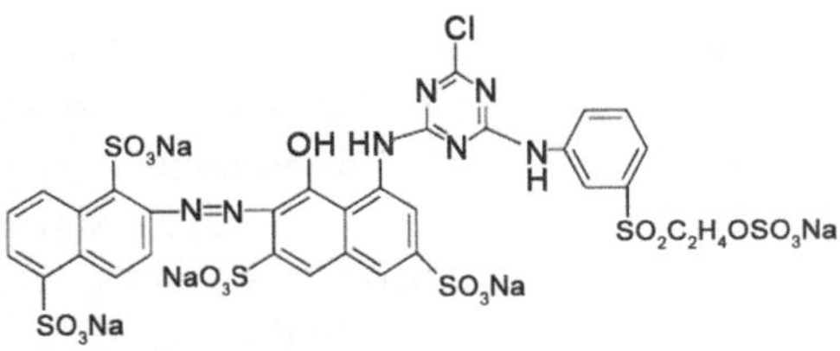

C.I. Reactive Red 195

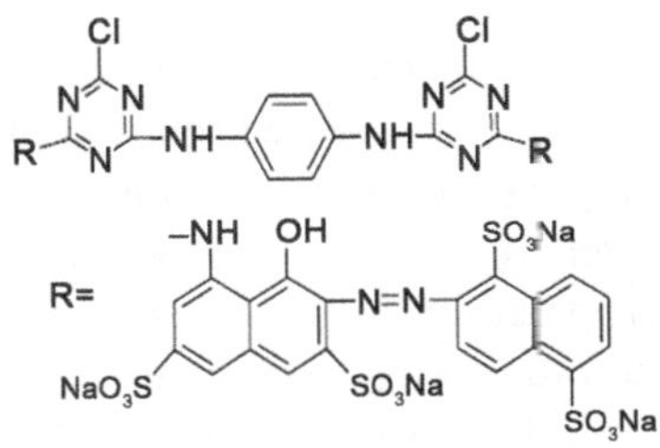

C.I. Reactive Red 141

Figure 21.1 Structures of two reactive dyes [12]

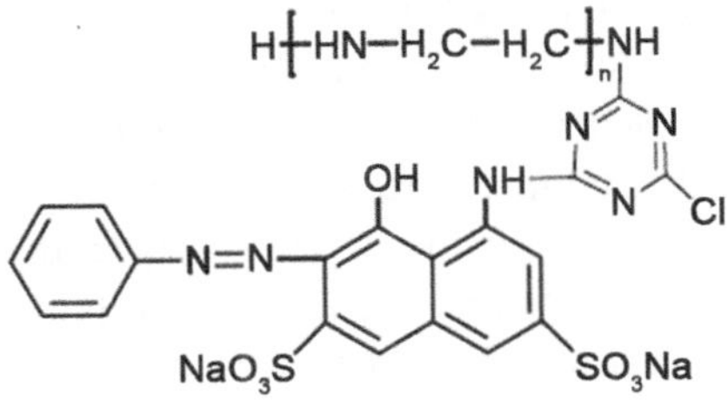

PAD1

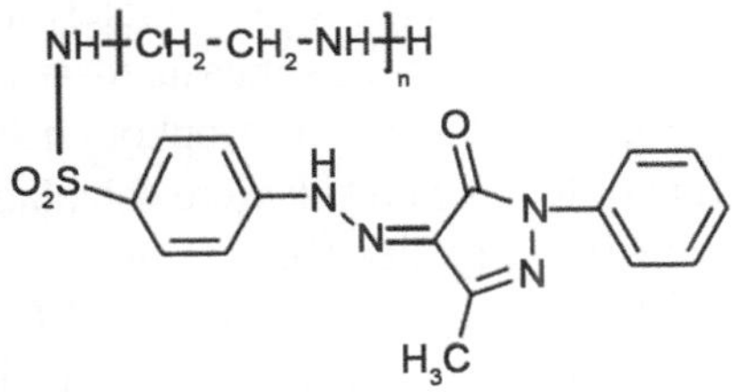

PAD2

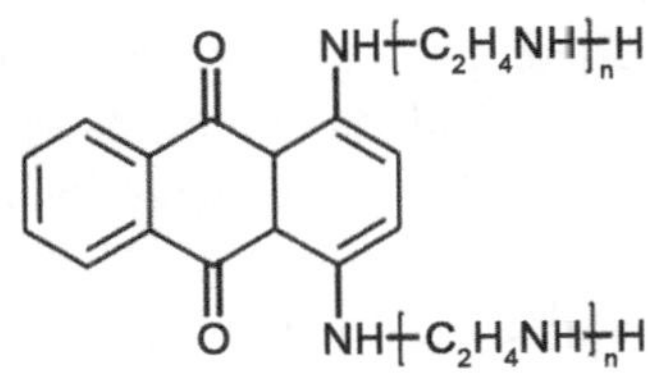

PAD3

Figure 21.2 Structures of cross linking dyes (n = ~ 4) [12]

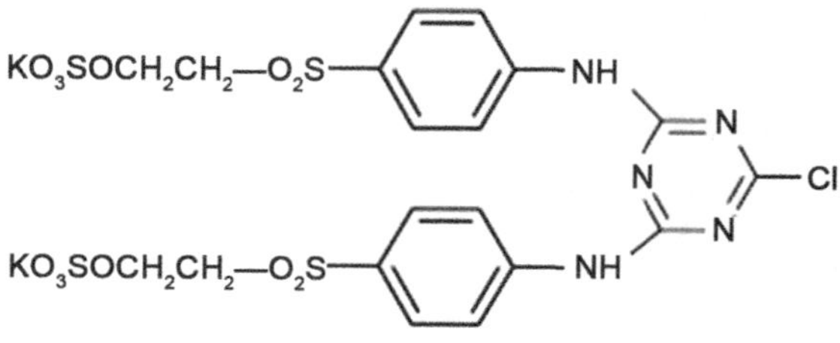

Figure 21.3 Structure of cross linker XLC

Silk fabrics were immersed into the dye-bath with dye concentration of 0.67 g/L, i.e. 2% owf (on weight of the fiber) at a liquor ratio of 30 : 1 at room temperature; 60 g/L sodium sulfate was added in batches to promote dyeing. After 10 min, the temperature was raised at 2°C/min to the dye fixation temperature: for C.I. Reactive Red 3, the fixation temperature was 40°C; for the other three reactive dyes, the temperature was 90°C. Sodium carbonate (2.5 g/L) was then added for dye fixation and the temperature was kept d from the dye-bath and washed thoroughly, and the rinse and the residue dye liquor were collected for measuring absorbance. The dyed silk fabrics were soaped with 5% OP-10 (alkyl phenol polyoxyethylene ether) solution at 95°C for 10 min, and then washed. The absorbance has been measured by collecting the soaping and washing residue liquors. Silk fabrics were immersed into the dye-bath with dye concentration of 0.67 g/L, i.e. 2% owf at a liquor ratio of 30:1 at room temperature, and then the temperature was raised to 60°C at the rate of 2°C/min and kept at that temperature for 1 h. After dyeing, fabrics were removed from the dyebath and air dried. Treatment with crosslinker XLC was as follows: the dyed silk fabrics were dipped into fixing-bath containing 20 g/L crosslinker XLC at a liquor-to-fiber ratio of 15 : 1, and a '2-dip-2-nip' operation was used at room temperature. The pressure on the mangle was adjusted to give approximately 80% wet pick-up, then the fabrics were heated to 60°C and kept over 10 min in an oven. Finally, the dyed fabrics were rinsed with water and soaped with 5% OP-10 solution (95°C) completely and air dried. The absorbance of the diluted dye solution was measured at the wavelength of maximum absorption (λ_{max}) of the dye using spectrophotometer. The dye exhaustion, reactivity and fixation of the dye on silk were calculated by use of equations [8]. The crease recovery properties were evaluated by measuring the dry wrinkle recovery angle (WRA) as per ISO standard. WRAs of the samples in warp (W) and fill (F) directions were measured separately; the recorded value of WRA (W+F) was the average of 10 measurements. The increment ratio of WRA (IRW) has been calculated. ASTM standards have been used to determine the fabric stiffness. The tensile strength of the sample has been determined according to ISO specification, and each value is the mean of the results from five tested samples. The color fastnesses to rubbing and washing have been tested according to ISO specifications.

21.3 Findings of the investigation

21.3.1 Anti-wrinkle properties of silk fabrics dyed with reactive dyes and crosslinking dyes

For the dyeing of the silk fabrics, four types of reactive dyes, namely C.I. Reactive Red 3, C.I. Reactive Red 2, C.I. Reactive Red 195 and C.I. Reactive Red 141, and three types of crosslinking dyes, namely PAD1, PAD2 and PAD3 have been utilized. Figures 21.1 and 21.2 depict the structures of these dyes. The dyeing properties, like exhaustion (*E*), reactivity (*R*) fixation (*F*), and color fastness of these dyes and the anti-wrinkle properties of the dyed silk fabrics have been studied. The exhaustions of the reactive dyes accounted to about 80%, and reactivities ranged between 88% and 98%. Exhaustions of PAD1, PAD2 and PAD3 were 74.8%, 96.6% and 99.6%, respectively. The reactivities of all three crosslinking dyes were above 99.0%. The differences of the exhaustions and reactivities for the dyes may be contributed to interaction between dyes and silk fibers. Under neutral dyeing conditions, silk fibers showed negative charge on the surface. A repulsion force existed between the silk and the reactive dyes and PAD1 due to the existence of sulfonic groups within the dye molecules. However, PAD2 and PAD3 showed certain positive charges and lower solubility under the same conditions, owing to the amino and imino groups in their molecules. Therefore, PAD2 and PAD3 exhibited higher substantivity to silk fibers, and thus higher dye exhaustion. In addition, the amino and imino groups of the crosslinking dyes were more reactive for the crosslinking reaction, and their reactivity towards the silk fibers was especially high, resulting in high dye fixation [8]. After the dyeing process, the rub fastness and wash fastness of these dyes reached 3–4 grade and 4–5 grade, respectively, which satisfied the application requirement. The WRA and IRW of the silk fabrics dyed with four different reactive dyes have been determined. Compared with the blank sample, the WRA of the silk fabric dyed with monofunctional reactive dye, C.I. Reactive Red 3, increased by 3.8%. However, when C.I. Reactive Red 2, which is generally regarded as a monofunctional reactive dye, was applied, the IRW was 11.1%. The increase of WRA might result from the crosslinking between C.I. Reactive Red 2 molecules and silk fibers caused by substitution of the two chlorines in part of C.I. Reactive Red 2 molecules by amino groups in the silk fibers. Interestingly, the IRWs of silk fabrics dyed with bifunctional reactive dyes,

C.I. Reactive Red 195 and C.I. Reactive Red 141, were 14.9% and 15.3%, respectively. These values were higher than those obtained from both C.I. Reactive Red 3 and C.I. Reactive Red 2. This improvement of the anti-wrinkle property was mainly attributed to the fact that the two bifunctional reactive dyes could react with silk fibers through their two reactive groups, which led to the crosslinking of silk fibers and limited the relative movement of the molecular chains. The IRWs of silk fabrics dyed with PAD1, PAD2 and PAD3 were between only 6.8–10.6% in the absence of XLC, which was close to those of C.I. Reactive Red 3 and C.I. Reactive Red 2. This increase in WRA presumably resulted from the deposition of the dyes on the fibers. However, when the crosslinker XLC was added, the WRAs of dyed silk fabrics were increased by 17.9%, 20.9% and 21.3%, respectively, for PAD1, PAD2 and PAD3, compared with that of the blank sample. These values were much larger than those of the bifunctional reactive dyes C.I. Reactive Red 195 and C.I. Reactive Red 141. Crosslinker XLC played an important role in improving the wrinkle resistance. Since there were three reactive groups which could react with crosslinking dyes or silk fibers within each molecule of crosslinker XLC, these reactive groups may have formed covalent bonds with both the crosslinking dyes and the silk fibers through several manners. Thereafter, the structure stability of the fiber has been improved, and the wrinkle resistance of the dyed silk fabric was enhanced. Compared with the selected finishing agents, the cross linking dyes in the presence of XLC exhibit comparable anti-wrinkle properties to BTCA, and a similar effect to EPSIB. The concentration of the crosslinking dyes has been 0.67 g/L and the applied XLC is only 20 g/L. Hence, it could be deduced that the crosslinking dyes can be applied to silk dyeing with satisfactory dyeing properties and be able to provide the silk with excellent anti-wrinkle properties as well.

21.4 Effect of the reactive dyes and the crosslinking dyes concentrations on anti-wrinkle properties

As the dye concentration increased from 0.67 g/L to 4.0 g/L, the WRAs of silk fabrics dyed with C.I. Reactive Red 2, C.I. Reactive Red 195 and C.I. Reactive Red 141 increased from 262°, 270° and 271° to 271°, 280° and 279°, respectively. The silk fabrics had IRW of about 4%. It indicates that the covalent bond formations between the dyes and fibers increase with the rise of dye concentration, and hence the crosslinking degree within the fibers would increase, leading to the improvement of wrinkle resistance. However, no significant change in WRA of silk fabrics was observed for further increase of the dye concentration to 6.0 g/L. As for the monofunctional dye-C.I. Reactive

Red 3, with the increase of the dye concentration from 0.67 g/L to 6.0 g/L, the WRA was improved by 5° (IRW was 2%). This increase was mainly due to deposition of an increasing number of dye molecule on fibers as the dye concentration increased. The effect of the concentration of the crosslinking dyes on WRA of silk fabrics with and without addition of crosslinker was examined. It was found that without being treated by crosslinker XLC, the silk fabrics dyed with PAD1, PAD2 and PAD3 also showed an increase of WRA by 10° with the increase of the dye concentration from 0.67 g/L to 6.0 g/L. Compared with the undyed silk, the IRW was 11–15%. This suggested that crosslinking dyes may deposit into the inside of the fibers, resulting in increasing the stability of the fiber structure. However, when 20 g/L of crosslinker XLC was used, the WRAs of the silk fabrics increased much more than those of one without being treated by crosslinker XLC as the dye concentrations increased. It was seen that as the concentration was increased from 0.67 g/L to 4.0 g/L, the WRA of silk fabrics dyed with PAD1 increased from 277° to 293° and the IRW increased from 17.9% to 24.7% with blank sample as a reference. Under the same conditions, the WRAs of silk fabrics dyed with PAD2 and PAD3 improved from 284° and 285° to 301° and 303°, respectively, and the IRWs reached 28.1% and 28.9%, respectively. The above results could totally achieve the finishing effect of EPSIB and glyoxal on the silk fabrics. It can also be observed that when the dye concentrations were further increased to 6.0 g/L, the change in WRA was insignificant. In fact, 4.0 g/L of dye concentration was sufficient for WRA improvement under the experimental conditions. Conclusions may be made according to the above results that both crosslinking and deposition mechanisms functioned to improve the WRA of silk fabrics dyed with the reactive dyes and the cross linking dyes; and the formation of the crosslinking was the dominant factor for the improvement of silk fabrics' WRA. Table 21.3 shows the effect of dye concentration on *E* (%), *R* (%) and *F* (%) of the dyes. It may be deduced that with the increase of the dye concentration from 0.67 g/L to 4.0 g/L, the exhaustion of both the reactive dyes and the crosslinking dyes decreased significantly. As for the reactive dyes, the decrease was 30–50% and it was 20–30% for the crosslinking dyes. The smaller reduction in exhaustion of the crosslinking dyes might be from their better substantivity to silk fiber. Furthermore, the results of reactivity in Table 21.3 indicated that the reactivity of the reactive dyes (82.9–98.4%) with the silk fibers was also lower than that of the crosslinking dyes (98.0–99.9%). Thus, the dye fixation of the crosslinking dyes has been perceptably greater, leading to a greater number of cross links between the dyes and the fibers. The WRA and IRW of the crosslinking dyes which were much larger with the increase of dye concentrations, could be ascribed to the relatively higher exhaustion and fixation of the cross linking dyes on silk.

21.5 Influence of the concentration of cross linker XLC on anti-wrinkle properties

As the XLC concentration has been increased to 40 g/L, the WRA of silk fabrics dyed with PAD1, PAD2 and PAD3 could rise up to 282–293° and the IRW was about 20.0–24.7% of that of the blank sample. Owing to the formation of abundant covalent bonds between the silk fibers and the dyes, two fiber molecules, and between two dye molecules, there can be an improvement of the wrinkle resistance of the dyed. When the concentration of cross linker was further increased to 100 g/L, the IRW was only increased by 3% compared with that obtained from application of 40 g/L agent. Therefore, usage of 40 g/L cross linker was sufficient for anti-wrinkle treatment of silk fabrics. It was also shown that when 0.67 g/L dye and 40 g/L XLC were used, the obtained WRA of silk fabrics was similar to that of silk fabrics treated by finishing agents, such as epoxide EPSIB and glyoxal. The findings revealed that the application of the crosslinking dyeing can simultaneously accomplish dyeing and finishing for silk fabrics without the need for finishing agent.

21.6 Influence of the concentrations of dyes and cross linker XLC on stiffness of silk fabrics

The handle behavior of the treated fabrics has been assessed by stiffness. The smaller the value of stiffness is, the softer the handle will be. It can be seen from Table 4 that as the dye concentration was increased from 0.67 g/L to 2.67 g/L, the stiffness of silk fabric samples increased from 23.1–23.7 mg·cm to 23.9–24.5 mg·cm, which was negligible. When the dye concentration was increased to 4.0 g/L, the stiffness of silk fabrics dyed with the reactive dyes and crosslinking dyes was further increased to 24.8–25.3 mg·cm. The fibers were slightly stiffer, but their handle was still satisfactory and met the requirement for application. Table 21.5 shows the stiffness results with the increasing addition of the cross linker XLC. When the concentration of XLC was 20 g/L, the stiffnesses of silk fabrics were 23.1, 23.7 and 23.3 mg·cm for the silk dyed PAD1, PAD2 and PAD3, respectively. Compared with the stiffness of the blank sample (23.1 mg·cm), the variation of handle of silk fabric was negligible. When the concentration of the cross linker was 40 g/L, the stiffness was 24.4–24.8 mg·cm, the change in the handle of silk fabric was insignificant. But, the dyed silk fabrics were a little stiffer as the concentration was higher than 40 g/L, particularly when 100 g/L of PAD2 and PAD3 have been used.

21.7 Tensile strength of the silk fabrics dyed with crosslinking dyes

The tensile strength retention of the silk fabric samples treated by finishing reagents or dyed with crosslinking dyes in the presence of XLC has been studied. The blank silk fabric sample was assumed to have 100% tensile strength retention. When silk fabrics were treated by finishing reagents BTCA and glyoxal, the tensile strength retention of samples decreased to 90.2% and 86.0%, respectively [5, 6]. The loss of tensile strength retention for the two samples was more than 9%. However, when using the crosslinking dyes assisted by the cross linker XLC to dye silk fabrics, the decrease of tensile strength of the samples was only 4–5%. The values were smaller than those of samples treated by the finishing reagents [12]. Addition of anti-wrinkle finishing reagents to silk fabrics usually damages tensile strength because of the formation of rigidity crosslink structure. A crosslinking structure may form to counteract the loss of strength of silk fabrics in the case of the samples dyed with crosslinking dyes with cross linker XLC.

References

1. Yang, Y. Q., and Li, S. Q., Dyeability of Silk Fabric Finished for Durable Press with Trimethylolmelamine, *Am. Dyest. Rep.* **82**(8), 22–28 (1993).

2. Schramm, C., and Rinderer, B. Multifunctional Carboxylic Acids in DP Finishing with BTCA and CA, *Textile Chem. Color. Am. Dyest. Rep.* **32**(4), 50–54 (2000).

3. Xu, W. L., Cui, W. G., Li, W. B., and Guo, W. Q. Two-step Durable Press Treatment of Cotton Fabric, *Color. Technol.* 117(6), 352–355 (2001).

4. Schramm, C., and Rinderer, B. Durable Press Finishing of Cotton with 1,2,3,4-Butanetetracaboxylic Acid and Glyoxal in a Two-step Process, *Cell. Chem. Technol.* **40**(1–2), 125–131 (2006).

5. Yang, Y. Q., and Li, S. Q. Crease Resistant Finishing of Silk Fabric with BTCA, *Textile Chem. Color.* **26**(5), 25–29 (1994).

6. Choi, H. M., Park M. J., and Oh, K. W. Nonformaldehyde Crease-resistant Finishing of Silk with Glyoxal, *Textile Chem.Color.* **30**(12), 41–45 (1998).

7. Cai, Z. S., Jiang, G. C., and Qiu, Y. P. Chemical Modification of Bombyx mori Silk with Epoxide EPSIB, *J. Appl. Polym. Sci.* **91** (6), 3579–3586 (2004).

8. Li, Y. L., Zhang, S. F., Yang, J. Z., Jiang, S., and Li, Q. Synthesis and Application of Novel Crosslinking Polyamine Dyes with Good Dyeing Performance, *Dyes Pigm.* **76**(2), 508–514 (2008).

9. Li, Y. L., Tang, Y. F., Zhang, S. F., and Yang, J. Z. Synthesis and Application of Crosslinking Blue Anthraquinone Polyamine Dye with High Fixation, *Textile Res. J.* **77**(9), 703–709 (2007).

10. Tang, B. T., Zhang, S. F., Yang, J. Z., and Liu, F. Synthesis of a Novel Water-soluble Crosslinking Polymeric Dye with Good Dyeing Properties, *Dyes Pigm.* **68**(1), 69–73 (2006).

11. Lei, X. P., Lewis, D. M., Shen, X. M., and Wang, Y. N. Crosslinking Nucleophilic Dyes on Wool, *Dyes Pigm.* **30**(4), 271–281 (1996).

12. Nan Yan Dang, Wei Ma, Shu Fen Zhang, Bing Tao Tang and Jin Zong Yang. Studies on Anti-wrinkle Properties of Silk Fabrics Dyed with Reactive and Crosslinking Dyes, Textile Research Journal **80**(4): 374–382.

22

Polyester/waste silk core spun yarns and fabrics

Summary : Owing to improper carding and fiber damages during staple fibre cutting, core-spun polyester/silk and 100% silk yarns have more irregularity and very high level of imperfections. There is very good improvement in yarn uniformity characteristics and tenacity after the single yarns are plied. Core-spun yarns having filament at the core and silk yarn fed left (z-twist) to the silk fibers during spinning undergoes wrapped ribbon twisting and poor sheath coverage. The core-spun yarns with center and right filament feeding have better sheath coverage. The filament rich core-spun yarns (3 and P3) have high tenacity and elongation due to the property of polyester. The strength realization from the core-spun yarns shows that the sheath fibers are also contributing to the tenacity of the yarns. Filament rich yarns have higher initial modulus and bending rigidity. The coefficient of friction (μ) of yarns increases as the polyester core content increases. Silk rich core-spun yarn (67% silk) has less hairiness, mass irregularity, thick places, and neps than the yarn with less silk content (33% silk). The thermal conductivity of silk rich fabrics is higher compared to polyester rich fabrics. The thermal resistance of the core-spun yarn fabrics decreases with the decrease in fabric thickness. The fabrics from core spun yarns predominant in silk show lesser water vapor permeability % than those for fabrics made from yarns predominant in polyester. The core-spun yarns predominant in polyester exhibits greater wicking than the fabrics predominant in silk. The fabrics predominant in polyester give unpleasant cooling effect than fabrics predominant in silk.

22.1 Introduction

It is an industrial practice to convert waste into useful products. Spun silk yarns are produced from staple fibers in industry from degummed non-reelable silk filaments degummed, cut into staples and processed through ring spinning system. Core-spun yarn with polyester as core and staple silk as sheath brings the advantages of durability of polyester and comfort characteristics of silk to the end products. Fabrics made from polyester core and waste silk sheath are stronger, having higher abrasion resistance than those from 100% silk yarn [1]. To produce core-spun yarns, filament yarn is fed in to the front roller nip through an attachment fitted to ring frame. Various types of attachments to feed core yarns are developed to produce different yarn structures and a few commercial systems are available. There is no literature relating to elaborate investigations on tensile, evenness, friction, stiffness, hairiness characteristics of polyester/silk core-spun yarns in relation to 100% polyester and 100%

DOI: 10.1201/9781003735427-22

silk yarns and comfort characteristics of fabrics made from these yarns viz., thermal, water vapor permeability, and wicking behavior; and hence an attempt has been made to address these aspects.

22.2 Pre-spinning operations on silk fibers

Mulberry silk waste fibers and polyester filament yarn have been used to conduct the investigation. The silk fibers have been processed two times in carding and then card slivers have been drawn through two passages. Silk roving of 2.4s has been produced from the finisher drawn slivers [8].

22.3 Production of core-spun yarns and plied yarns

A core spinning attachment has been designed to produce polyester/silk core-spun yarns and fixed on a ring frame. The tension of the filament yarn has been controlled by a disc tensioner before feeding into the nip of front drafting rollers. For the core material, a 50-denier flat polyester filament with 48 filaments has been chosen. During the feeding at the front roller nip, the tension on the core has been maintained at 10% of the filament yarn breaking load. Single, double, and triple ends of filament yarn with various core-sheath ratios of 33/67, 67/33, and 100/0%, have been used as core component. Yarns of 34s Ne (nominal) were produced [8]. To investigate the effect of core positioning on sheath coverage over core, core-spun yarns of 33/67 polyester/silk were produced wherein colored polyester yarn was fed to the front roller at three positions, viz. right, and left and at center with respect to drafted silk fibers. The core component was dyed at 4% shade with disperse navy blue before feeding. After producing the core-spun yarns with different core positioning, the yarns were wound on wrapping board for visual inspection. From these wrappings and through Leica projection microscope, it was found that coverage of core component by silk fibers was the highest for yarns where core was positioned at right of the drafted silk fibers followed by yarns produced with center and left core positioning. The core positioning photographs were shown in Figures 22.1(a), 22.1(b), and 22.1(c), respectively. Consequently for all the experiments, right-core positioning was selected. Z-twisted yarns were produced at ring frame with a spindle speed of 10,000 rpm, front and back-zone settings of 44 mm and 60 mm, respectively, break draft of 1.14 twist per inch of 16 and a traveler size of 3/O (UIULUDR). The other production parameters for all the single yarns have been determined. The ratio of ply twist (tpi) to single twist was kept 0.80.

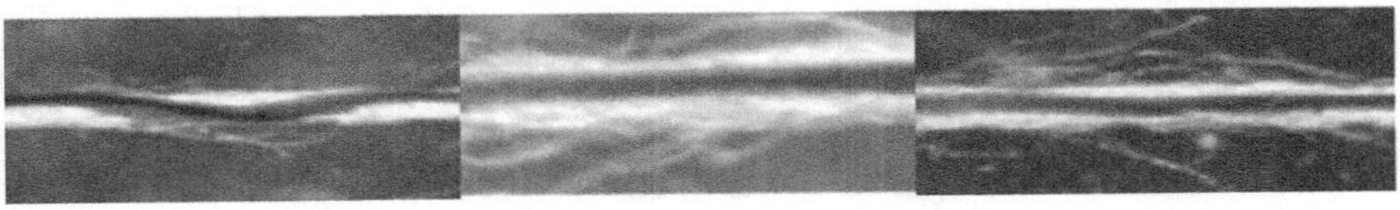

(i) (ii) (iii)

Figure 22.1 (i) Left core, (ii) Centre core, (iii) Right core [8]

Two ends of each of these yarns 1–4 were separately assembly wound and then ply-twisted in S-direction in a twisting machine. The two plied yarns made from these were coded as P1–P4. The ply twist was kept at 13 tpi. A 1400 gauge single jersey knitting machine has been used to knit the plied yarns after steaming and washing. Four different fabrics have been produced from the single yarns (1–4).

22.4 Testing aspects

The staple silk fibres from sliver have been tested for fineness and length. Also the yarn diameter has been measured. The fabrics produced have been tested for thickness, comfort, relative water vapor permeability, and wicking behavior [8].

22.5 Influence of core positioning on structure and properties of core-spun yarn

The fixing of the core component in relation to drafted fibers at the front roller nip during yarn formation influences the coverage of sheath fibers around core component in the core-spun yarn [8]. Many photographs of the Z-twisted core-spun yarns (33/67%) made from colored-core positioned at left (L), center (C), and right (R) of the drafted silk fibers at 5 cm interval are taken. They clearly point out that core-spun yarn with core situated on right and center has better coverage of the core component than the yarn made from core situated on left. The core-spun yarn made with polyester core positioned at left has a wrapped ribbon-like structure where both the polyester filaments and staple fibers are ribbon-twisted together. While twisting fibers in Z-direction as the fibers emerges from front roller nip, fibers or filaments from the left side of the spinning triangle encounter friction at top breast of the bottom front drafting roller. This delays the twist and tension transfers to this part of spinning triangle. Consequently, the fibers at left side of the spinning triangle are slack and join the forming yarn later than those fibers at center and right side of the spinning triangle. Due to this, the drafted silk fibers at center and right while emerging out are partly twisted before joining the core filaments from left leading to insufficient coverage of the polyester filaments. The tenacity of core-spun yarn produced with core positioned at center of spinning triangle is lower than the yarns produced with left and right core position. This may be due to more obliquity of the filaments with center fed core as the filaments are twisted with respect to an axis that is not helical. It is evident from Figure 22.1(b) that polyester core is uniform in diameter and is at center of core-spun yarn indicating that core has undergone cylindrical twisting. With analyzing several photographs of core-spun yarns, it is found that core polyester has mixture of cylindrical (in majority cases) and ribbon twisting (few cases) along the yarn in the case of yarns where core is positioned at right. On the other hand, the core-spun yarn produced with core positioned at left has mostly of ribbon twisting evidenced from kinks and varying diameter of core component. Hence, the core-spun yarns with left and right core positioning are subjected to varying radial pressure during tensile testing resulting in readjustment of fibre and filaments within the core-spun yarn and as a result have more CV% of breaking elongation. There is a space of about 2 mm between filament yarn axes to staple fiber ribbon axis at front roller nip.

22.6 Influence of core and sheath ratio on yarn properties

22.6.1 Yarn diameters

The different characteristics of yarns have been determined. There is increase in yarn diameter with increase in percentage of sheath which indicates difficulties in packing the silk fibers of different configurations in the yarn matrix. Hence, 100% silk yarn 1 is bulkier than the other yarns [8].

22.6.2 Yarn evenness characteristics

There is considerable improvement in mass irregularity characteristics, particularly thin places, due to doubling effect, after plying. The single as well as plied silk yarns (100%), ('1' and 'P1') and other core-spun yarns are more irregular with greater level of imperfections than the commercial cotton yarns [8]. The silk fibres have been carded on an old machine. Poor carding of the waste silk fibers, fiber damages, and wide distribution of fiber lengths are the reasons for very high imperfections and mass irregularity of 100% silk and polyester/silk core-spun yarns. The single and plied core-spun yarns ('2' and 'P2') with a core-sheath ratio of 33/67% have less CV% and fewer imperfections (except thin places) than the other yarn, 3 and P3 (67/33%) due to the better and even wrapping by more available silk fibers over the polyester filaments even though the yarn 3 has more number of uniform polyester filaments in the core. Increase in the polyester content is found to reduce the thin places.

22.7 Tensile characteristics of 100% silk and core-spun yarns

The plied yarns exhibit enhancement in tenacity, elongation, and CV%. Yarn '1' (100% silk yarn) shows greater CV% of tenacity and elongation in comparison with the core-spun yarns owing to its very high mass irregularity. With the increase in core %, there is enhancement in tenacity and elongation at break in the case of core-spun yarns due to strong and high extensible polyester core. The yarn '3' exhibits high elongation % (43.0) due to the use of two ends of polyester yarn in the core with lesser sheath %. The three ends polyester ring twisted yarns '4' show higher tenacity (36.9 cN/tex) with 33% elongation, which is higher, than the parent 50-denier filament yarn (tenacity of 35.6 cN/tex with 26.50% elongation.). The 50-denier yarn has percentage extension at break of 26.5 and a high CV% of extension at break (17.5) with non-catastrophic failure [8]. Cohesion among the filaments brought out by twist enhances the breaking extension and hence slight improvement in the tenacity of the yarn, 4 [2, 3]. Increase in the polyester content causes increase in the initial moduli of the yarns because of high stiffness of polyester compared with silk fibers. There is reduction in initial moduli after plying because of reduction in fiber interlocking since the single yarns get untwisted during plying.

22.8 Strength realization of core filament

22.8.1 Yarn in core-spun yarns

The strength realization of core component has been calculated as the ratio of breaking strength of core-spun yarn to breaking strength of core component x 100, under the assumption that core only contributes to the strength of the core-spun yarns. A strength realization of more than 100% for all the core-spun yarns shows that the sheath silk fibers have also contributed to tensile load. The contribution of sheath towards tensile load in the case of yarns with core-sheath ratio of 33/67% (2 and P2), is very high because of better wrapping (tight and even) of more number of silk fibers than the core-spun yarn 3, where silk fibers are less in number on the sheath. Though the mean fiber length of waste silk is around 25.5 mm, there are a few longer fibers that might have played a significant role in enhancing the frictional contact thereby increasing the strength realization. This indicates that sufficient number of sheath fibers is needed to enhance radial pressure over filaments and among staple fiber themselves to contribute strength to core-spun yarns [8]. Due to more alignment of the surface fibers along the ply yarn axis for plied yarn than the single yarn, the strength realization from plied yarns is higher than single yarns.

22.8.2 Frictional characteristics of yarns

In the case of plied yarns, the coefficient of friction is greater compared with the single core spun yarns and increases with increase in the polyester core content. The yarn diameter decreases as the polyester content increases which lead to reduce area of contact during rubbing. The higher friction values observed for polyester rich core-spun yarn, 100% polyester yarn and plied yarns are due to their high bending rigidities [8]. Rigid yarn is subjected to greater tension as it bends over metal surface leading to more output tension, due to greater friction.

22.8.3 Hairiness of core-spun yarns

Yarn 2 (33/67) has least number of hairs compared with other spun yarns [8]. There are more hairs on Yarn 1 compared with Yarn 2, because of the presence of more number of staple fibers in its cross section. When the sheath content reduces from 67% to 33%, again hairiness increases. It arises from poor and loose wrapping of fibers, during less availability of staple fibers as discussed earlier.

22.9 Fabric properties

22.9.1 Fabric thickness and cover factor

The thickness and cover factor of the fabrics have been determined. Because of bulkiness of silk fibers, fabrics made from 100% silk yarn (FS1) and silk rich yarns (FS2 and FS3) are thicker and having more cover factor than the fabrics made from 100% polyester filament yarn [8].

22.10 Thermal properties of fabrics

Due to the increase in the polyester core content, the thermal conductivity of fabrics reduces. It is because of the higher thermal conductivity values (both longitudinal and transverse) of silk fibers than polyester fibers [4, 5]. With the increase in core content, the thermal resistivity of the fabrics reduces, because of the reduction in thickness of fabrics. The peak heat flow represents transient behavior and it increases with increasing content of polyester core. The peak heat flow may also be influenced by specific heat of fibers used in the fabrics. The specific heat of dry silk and polyester fibers is 1.34 and 1.38 Jg/K, which are nearly same (physical properties of fibers, Hearle, p. 587). Transient thermal behavior of fabrics indicates warm or cool effect of fabrics and depends on the contact interface between skin and fabric [6]. Structure of fibers, yarns, fabrics, and thickness of outer layer of fabrics influence this behavior [6, 7]. Thinner and harder fabrics produced from 100% polyester yarns and polyester rich core-spun yarns have thin outer layer. Fabrics made from 100% silk and silk rich core-spun yarns are fuzzier with thick outer layer [8]. Also the silk fibers are triangular shaped, rendering silk and silk rich fabrics to a have less contact area to the measuring head or skin. These factors contribute to greater peak heat flow seen in the case of polyester rich fabrics compared to silk rich fabrics, and the former gives an unpleasant cool feeling to the wearer.

22.11 Water vapor permeability of fabrics

An increase in the polyester content causes increase in the water vapor permeability%. The polyester fabrics produce an unpleasant cool influence on the skin, as they absorb sweat or water, and transmit them to the near areas. However, silk materials absorb water or sweat because of its hydrophilic nature and the moisture retention depends on the thickness of the fabrics. The wearer experiences a dry comfortable feeling due to the hydrophilic nature of silk fibers coupled with high thickness of silk rich fabrics [8].

22.12 Wicking behavior of fabrics

Figure 22.2 depicts the increase in fabric weight due to wicking of fabrics. 100% polyester and polyester-dominant core spun fabrics exhibit greater wicking compared with other fabrics. This is due to filaments prone to wicking as a result of capillary action than the staple fibers. Further the thin 100% polyester fabric aid better wicking compared to other fabrics. In silk predominant fabrics, water passes through a difficult path between fibers of numerous configurations spaced loosely and leads to poor wicking [8].

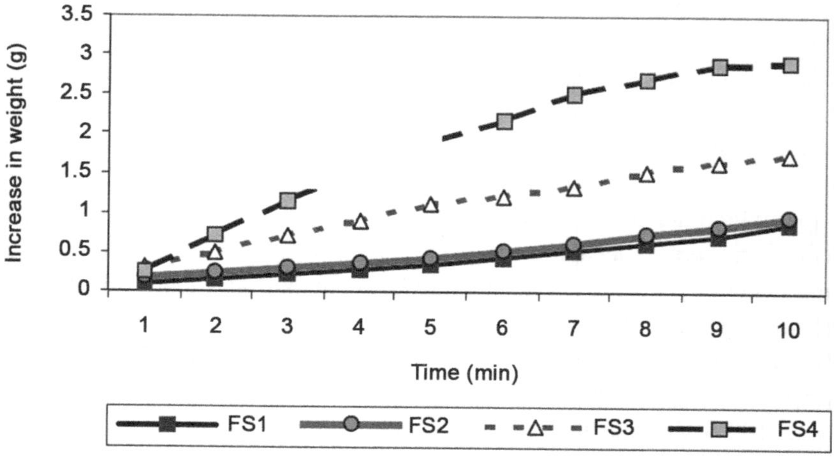

Figure 22.2 Wicking behavior of fabrics [8]

References

1. Vasumathi, B.V., Prabhu, J., Bhat, D. and Somashekar, T.H. (1994). Comparison of Silk/Polyester Core and Spun Silk Yarns and Fabrics, Indian J. Fibre Textile Res., 19: 247–250.
2. Rengasamy, R.S., Kothari, V.K. and Patnaik, A. (2004). Stress-strain Characteristics of Air-jet Textured Yarns, Indian J. Fibre Textile Res., 29: 283–289.
3. Hearle, J.W.S., Grossberg, P., and Backer, S. (1969). Structural Mechanics of Fibres, Yarns and Fabrics, Wiley Interscience, London.
4. Kawabata, S. and Rengasamy, R.S. (2002). Thermal Conductivity of Unidirectional Fibre Composites Made from Yarns and Computation of Thermal Conductivity of Yarns, Indian J. Fibre Textile Res., 27: 217–223.
5. Rengasamy, R.S. and Kawabata, S. (2002). Computation of Thermal Conductivity of Fibre from Thermal Conductivity of Twisted Yarn, Indian J. Fibre Textile Res., 27: 342–345.
6. Pac, M.J., Bueno, M.A., Renner, M. and El-Kasmi, S. (2001). Warm-cool Feeling Relative to Tribological Properties of Fabrics, Textile Res. J., 71: 806–812.
7. Schneider, A.M. and Holcome, B.V. (1991). Properties Influencing Coolness to the Touch of Fabrics, Textile Res. J., 61: 488–494.
8. Ramesh Kumar C, Rengasamy RS, and Anbumani N. Studies on Polyester/Waste Silk Core-spun Yarns and Fabrics, Journal of Industrial Textiles, 38(3), 2009, 191.

23

Relation between cocoon parameters and raw silk uniformity characteristics

Summary : Cocoon quality parameters have an important bearing on the quality of raw silk reeled. Earlier studies have addressed many of the factors but the influence of gradient of denier variation within a cocoon on uniformity characteristics of raw silk has not been specifically studied. In this work, eleven lots of cocoons of different races were taken and studied for its quality parameters including the slope of the denier curve. Later these cocoons were reeled on both multi-end and automatic reeling machines. The raw silk was tested for its average size, size deviation, maximum deviation, evenness variation and Uster U%. Correlation and regression analysis to study the strength of relationship between the cocoon parameters and raw silk uniformity characteristics were carried out. It was observed that the slope/gradient of the filament denier within a cocoon in addition to other cocoon parameters has significant influence on the raw silk uniformity characteristics.

23.1 Introduction

Raw silk quality is known to be largely depending on the cocoon properties [1], and cocoons, being a biological material, have many inherent variations. The cocoon properties are defined by a large number of parameters, some of which are important for the parent cocoons, for race maintenance, while some are important for reeling cocoons [2, 3]. For a reeling technologist, the technological parameters of the cocoon are significantly important, since they determine the quality, quantity and efficiency of the reeling process. The fact that the raw material, viz. cocoons, has a significant bearing on the process and the end product quality, with the added dimension that they are a highly variable commodity, makes the situation quite complex. In most studies [4–6], the influence of parameters of dropping ends and feeding ends and a few important properties of cocoons on raw silk size properties from a stochastic viewpoint have been considered. With the introduction of a large number of newer races of cocoons [7], it is observed that the filament denier follows different patterns, some exhibiting a steep fall, while some exhibit a flatter drop in the filament denier from the outermost to the innermost layers of the cocoon. It is therefore pertinent to study the impact of this aspect of cocoons on raw silk quality. Raw silk is expressed in terms of around eleven parameters.

DOI: 10.1201/9781003735427-23

A set of few of these parameters reflect the uniformity characteristics, a few of them influence the production efficiency in the later process, a few indicate the dynamometric properties, while yet another set of parameters refer to the appearance and cleanness characteristics. Raw silk denier is a highly variable function, since it is made from a raw material that has inherent variations within the cocoon as well as between the cocoons. In addition the component of dropping ends, where the continuity of the same filament in the raw silk thread is not ensured, adds to the variation. Earlier studies have attempted to establish the influence of cocoon parameters on each of these categories of raw silk quality. Shimazaki [8] has discussed in detail the influence of cocoon filament length and denier on the uniformity characteristics of raw silk reeled. He has provided a theoretical as well as practical approach to the findings in his study, where fixed size and fixed number techniques of reeling have been studied. Bai [9] has analyzed the influence of cocoon properties on raw silk uniformity characteristics, wherein a purely theoretical approach has been considered in establishing the relationship in the case of fixed size reeling.

Shimazaki had indicated that cocoon properties have a greater influence on size deviation, while the skill of the reeler has influence on evenness. The conventional method of evenness testing has now been changed from the average percent system to one of considering the frequency and intensity of the denier variation. But as to how the gradient of the cocoon filament denier from the outermost to the innermost layer would influence the uniformity characteristics of raw silk has not been specifically studied so far. Therefore the focus in this work has been to first determine the gradient of the denier curve of the different races and later to study the influence of this gradient on the uniformity characteristics of raw silk.

23.2 Raw silk uniformity characteristics

Raw silk uniformity characteristics can be classified under two heads, namely, the long-term denier variation which is expressed in terms of size deviation and maximum deviation, and the short-term denier variation which is expressed in terms of evenness. These tests have been a part of the classical methods of raw silk testing and grading. One more test for raw silk, of relatively recent origin, is the uster evenness test. This gives the overall variation in the fineness of raw silk, of relatively recent origin, is the Uster evenness test. This gives the overall variation in the fineness of raw silk expressed in terms of U%. This may indicate the short-term or the long-term variation, depending on the choice of test length. But the uster U% provides information on the overall variation that includes within and between variation. Within variation corresponds to the length of the transducer used in the tester. [10]

23.3 Cocoon parameters

The average results of filament denier, filament length, along with the standard deviation of filament denier within cocoons and between cocoons, have been obtained. The linear fit for estimating the slope of the filament denier curve was found to be a significant fit in all the cases. The graphs of the cocoon filament are shown in Figure 23.1, and Figure 23.2 shows the graph of the filament denier variation along the length of the cocoon of lot 1. [10]

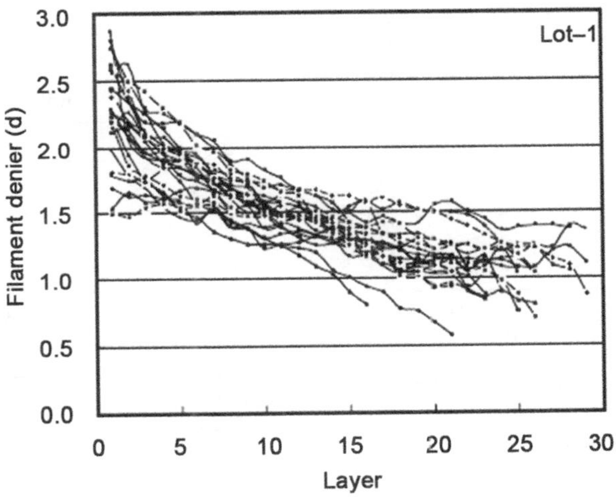

Figure 23.1 Cocoon filament size curves [10]

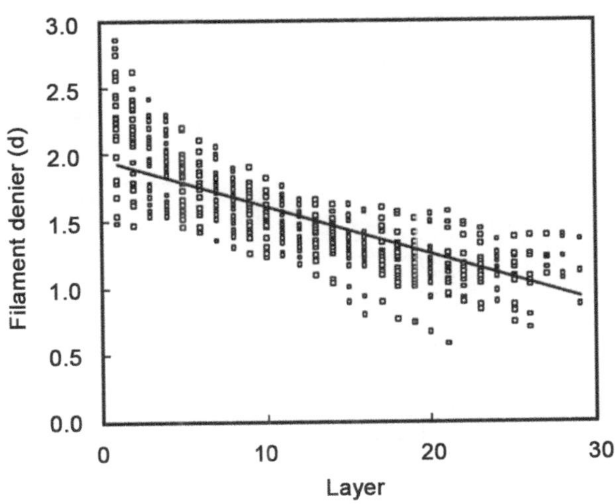

Figure 23.2 Linear curve fit for the filament denier curve [10]

This graph indicates the range of values of all the cocoons within a lot, along with the linear curve fit to determine its slope. Figure 23.3 shows the average change of cocoon filament size and the linear curve fit to determine its slope.

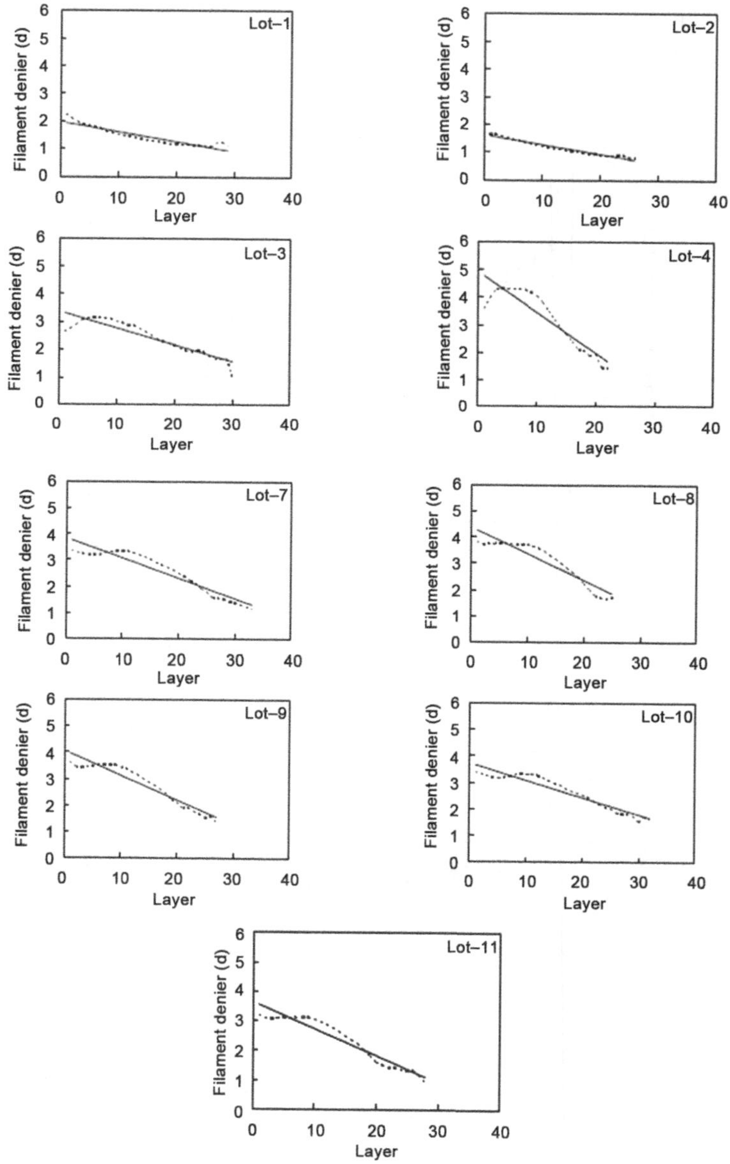

Figure 23.3 Linear curve fit for the filament denier curve for each lot [10]

ANOVA was run to test the significant difference in the slopes of the different lots of cocoons, and it was found that there exists a significant difference between the slopes of the different lots of cocoons. The values of the rate of change of filament length denoted as the gradient in respect of all the eleven lots of cocoons are depicted in Figure 23.4.

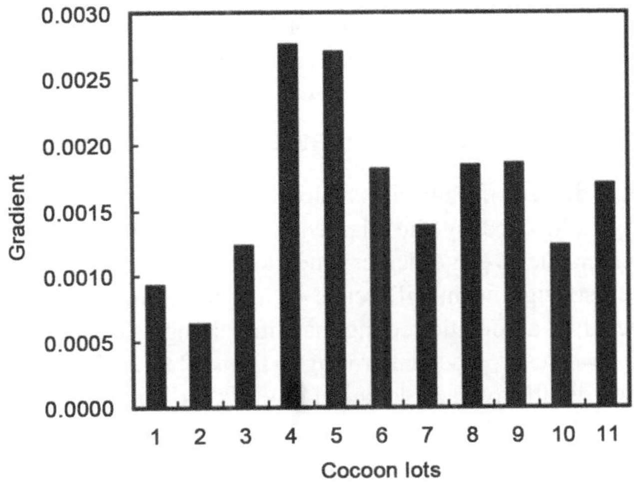

Figure 23.4 Gradient of cocoon filament denier [10]

ANOVA was run to test the significant difference in this parameter for different lots of cocoons and it was found that there exists a significant difference between the different lots of cocoons. It was also observed that this parameter has significant correlation with the slope parameter. The results of the size uniformity characteristics of raw silk expressed in terms of size deviation, maximum deviation, evenness variation I and II and uster U% in respect of raw silk reeled on automatic reeling machine and multi-end reeling machine were compiled and used for the statistical analysis.

23.4 Correlation and regression analysis

Correlation analysis was carried out between the cocoon parameters and the raw silk characteristics, separately for raw silk reeled on automatic reeling machine and multi-end reeling machine. Linear regression analysis was carried out in order to predict the value of the dependent variable from a set of the independent variables. Stepwise regression method was adopted, which would pick up the most significant variables from among the set of independent variables that best explain the variations in the dependent variable. It may be noted that in the case of evenness variation II in respect of both automatic reeling and multi-end reeling, no variables from among the independent variables got entered into the model. Similarly, in the case of average size in respect of multi-end reeling machine, no independent variable got entered into the regression model. Further the cocoon parameters of slope and gradient were able to explain a significant part of the variation in respect of most of the dependent variables. The variables picked up by the stepwise regression, along with the R^2 value of the model for each of the dependent variable, have been calculated. [10]

23.5 Average denier of raw silk vs. cocoon parameters

The average denier of raw silk reeled on automatic reeling machine is significantly influenced by the filament denier of the cocoon. Coarser the filament denier, the average denier tends to be higher. It was also observed that there exists significant difference in the denier of the different lots of raw silk reeled on automatic reeling machine, though the same type (22d) of size detector was used. The denier ranged from 22 to 24.5, while in the case of multi-end silk, the average denier of raw silk did not depend on any of the cocoon parameters. When the size controller picks up a cocoon, it cannot have a choice of cocoon from the outer, middle or inner layers as in the case of hand feeding. The probability of picking up a filament coarser than what is required, therefore, is higher in the case of a size controller. The higher the filament denier, the greater will be the difference in the denier, between what is actually needed and what is actually fed. This is perhaps reflected in the significant correlation of average denier with that of filament denier only in the case of automatic machine reeled silk and not multi-end. [10]

Shimazaki has discussed this aspect and mentioned that it is not necessary in the case of reeling with size controller that the ratio of filament size to the objective size of raw silk is an integral number and has given the formula for estimating the size of raw silk.

This clearly indicates the dependence of mean size of raw silk on the average size of feeding filaments. He further states that the limit size, which indicates the feeding point of a cocoon filament should be based on this formula for each given material cocoon lot.

23.6 Size deviation and maximum deviation of raw silk vs. cocoon parameters

The size deviation of raw silk, in the case of both automatic machine reeling and multi-end reeling, is significantly dependent upon the filament slope, filament gradient, filament denier and within cocoon size variation. There is no relation with average filament length. But in the case of multi-end, the size deviation, in addition, is related to between cocoon variation, i.e. heterogeneity within the lot. When the reeler finds a cocoon, the estimate of denier from outer, middle and inner layers would be almost a constant value in the mind. The difference that exists between the cocoons is not comprehendible, and the chances of error due to this difference, assume greater proportion here. While in the case of automatic reeling, the choice of cocoon feeding is a random occurrence. [10]

This has also been corroborated by the findings of Shimazaki [1], where he relates the size deviation of raw silk reeled on automatic reeling machine as a function of mean size feeding filament and the coefficient of size variance of feeding filament. While in the case of fixed number of cocoons, the size deviation of raw silk is defined as a function of mean size of cocoon filament and the standard deviation of the mean number of reeling cocoons. The maximum deviation in respect of multi-end reeling exhibits the same relationship as in the case of standard deviation. The cocoon parameters that influence size deviation influence maximum deviation too, almost in an identical way. Whereas in the case of automatic machine reeled silk, the maximum deviation is influenced by two additional cocoon parameters, namely average filament length and between cocoon size variation, while the other parameters that influence size deviation continues to influence the maximum deviation too. Maximum deviation is the extreme case of denier variations or to say the largest spring in the size both on the finer as well as well as the coarser side. It was also observed that the larger deviation had occurred in 9 out of the 11 lots towards the coarser side in the case of automatic machine reeled silk. That means that the denier had erred on the thicker side. While the phenomenon was exactly the reverse in the case of multi-end reeled silk. The maximum deviation, in 8 out of 11 lots, was towards the finer side. This error on the coarser side in the case of automatic machine reeled silk is actuated by lower filament length and higher between cocoon size variation.

23.7 Evenness variation of raw silk vs. cocoon parameters

In respect of evenness variation I, the occurrence of the denier variation is influenced by five cocoon parameters, viz. slope, gradient, filament denier, within cocoon size variation and between cocoon size variations. Evenness variation I are the stripes that are caused by a raw silk denier variation of say up to around 6.5% of the average value and in this case a little more than one denier on either side. Anything, more than this, would get classified as evenness variation II. For the estimator to comprehend the stripe, and take a count of it, the variation has to be at least for around a length of 10 m, which would form a band width of, say, 3 mm. If the variation is across shorter lengths than this, it doesn't get reckoned as a stripe under evenness variation. Thus denier variations over such lengths in a range of 1 denier on either side would be more due to the cocoon parameters. It cannot be caused due to the difference in the number of feeding ends. Denier variations of around 13% from the average value are classified as evenness variation II. Thus evenness variation II are stripes of variations in denier over and above 1 denier and perhaps around 3 denier in this case, and the length of stripe corresponding to the explanation provided under evenness variation I. It is seen that these variations are not caused by any of the cocoon parameters. Since such differences can occur due to the difference in the number of feeding ends. It appears that they are caused more due to the delay in feeding the cocoon or feeding more than what is required and clearly due to the skill of the reeler in the case of multi-end reeling, while they are more due to random occurrences of the delay or overfeeding in the case of automatic reeling. [10]

23.8 Uster U% of raw silk vs. cocoon parameters

Uster test can detect denier variation within as well as between, over short ranges of even a few cm and the U% is significantly explained by the slope, gradient, filament length and filament denier of the cocoons. The trend of relationship has been similar in both the methods of reeling. Uster U% would be accumulated value of the unevenness, which can be within interval of a few cm up to 400 m. Obviously, the component of short length variations can be caused more by the inherent cocoon variations. The dependence of Uster U% on the filament length reiterates the observation of Shimazaki [8], that the thread length, which is continuously composed of same cocoon filaments has influence on the evenness of raw silk. The evenness that is referred to here is the overall evenness and not a categorization of frequency and the intensities of denier variation. This overall evenness is more akin to the Uster U%. The continuity of thread length of same cocoon filaments is more related to its non-broken filament length and in turn the filament length. [10]

23.9 Comparison of actual size deviation with that of theoretical approach

Shimazaki [1] while discussing the relation between the size property of cocoon filament and the size property of raw silk has derived the formula for estimating the size deviation of raw silk reeled on automatic reeling machine. The estimated size deviation of raw silk for each of the 11 lots of cocoons was thus calculated using the formula put forth by him. The estimated size deviation was later compared with actual size deviation in respect of all the 11 lots (Figure 23.5). A great degree of agreement between the two values is observed. [10]

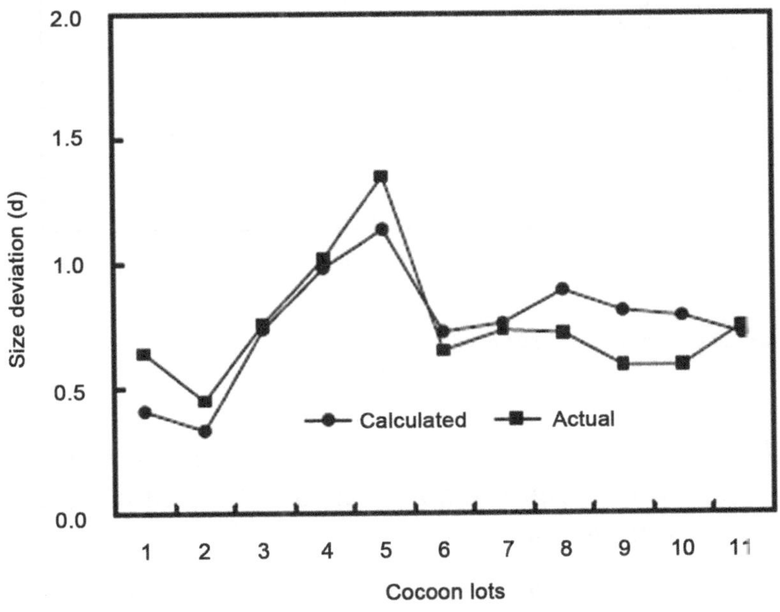

Figure 23.5 Comparison of calculated and actual size [10]

Similarly the formula proposed by Shimazaki [1] for estimating the size deviation of raw silk in the case of reeling with fixed number of cocoons was applied in the present case. Using the formula (3), an estimate of the size deviation of raw silk reeled on multi-end reeling machine based on the cocoon parameters was made and compared with the actual (Fig. 23.6). It may be observed that they are in close agreement, except in the case of two lots of fine denier cocoons. It may be observed that they are in close agreement, except in the case of two lots of fine denier cocoons. Theoretically, the fine denier cocoons with their low size variance are capable of providing raw silk with much lower size variation. But the manual limitations of maintaining a large number of cocoons per end take the size deviation to a slightly higher level. It may be noted that this size deviation, however, still continues to be better than in the other cases.

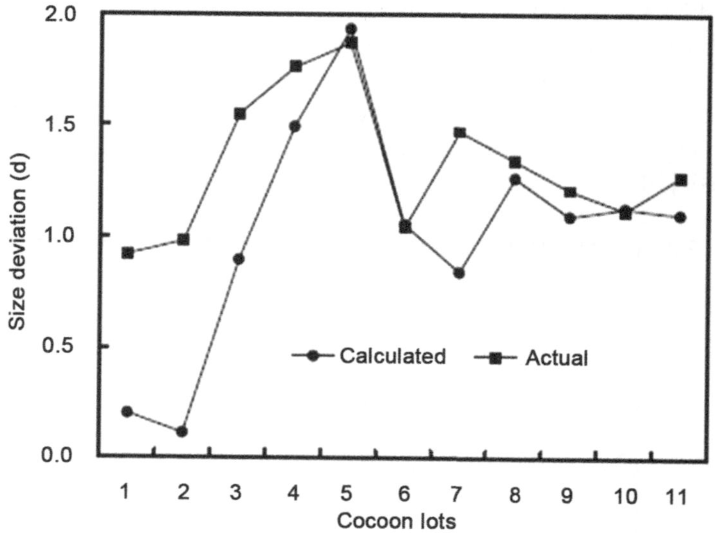

Figure 23.6 Comparison of calculated and actual size deviation of raw silk reeled on multi-end reeling machine [10]

23.10 Comparison of the uniformity characteristics of raw silk reeled on automatic and multi-end reeling machine

The size deviation, evenness variation I, evenness variation II, and Uster U% of raw silk reeled on automatic reeling machine and multi-end reeling machine were compared (Fig. 23.7 to 23.10). It is observed that as far as long-term variation is concerned, i.e. size deviation, the performance of automatic reeling machine is better, while in the case of short-term variation multi-end has outperformed automatic reeling machine, which is quite surprising [9]. With highly skilled reelers, the maintenance of right number of cocoons throughout the reeling time is better ensured. While the long-term variation which is due to cocoon variation is better managed by the automatic reeling, since it purely acts when limit size of cocoon feed is reached, irrespective of the number of cocoons from the outer, middle or inner layers that are being reeled out. The lower efficiency of automatic reeled silk, with regard to evenness, is very well explained by Shimazaki [8]. He states that the control of evenness is how less we manage the delay from the requirement of feeding filament from the reeling end to finish feeding. He further states that in reeling with the size controller, the effect of controlling the delay is less than the effect expected in reeling with fixed number of cocoons and the efficiency of evenness defect may be about 0.2 to 0.4, in this case in comparison to the fixed number reeling. These two short-term variations, namely evenness variation I and Uster U% are found to depend on the cocoon parameters as well as the skill of the reeler. As regards the evenness variation II, there is no difference between the two methods of reeling. As can be seen from the analysis carried out, that evenness variation II depends only on the skill of the reeler and has got nothing to do with the cocoon parameters. With the skill level of the reelers in the case of multi-end, matching the efficiency of automatic reeling, no observable difference has been found. [10]

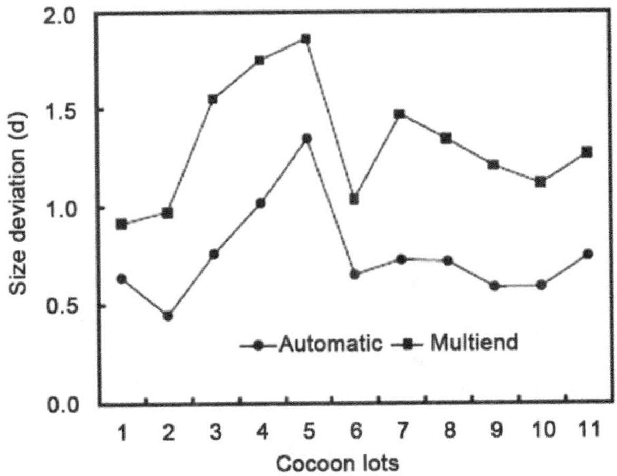

Figure 23.7 Comparison of size deviation of raw silk [10]

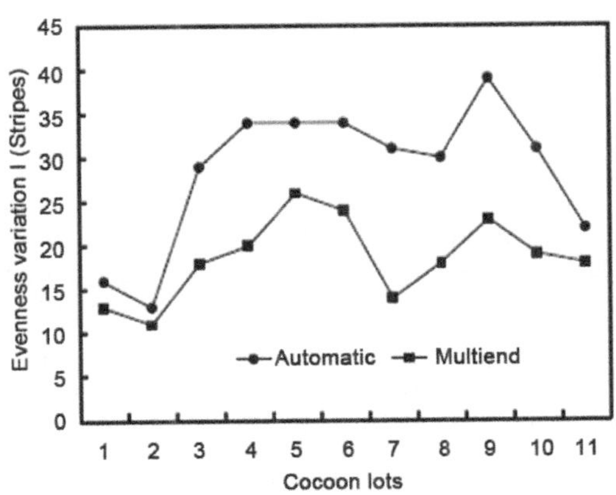

Figure 23.8 Comparison of evenness variation I of raw silk [10]

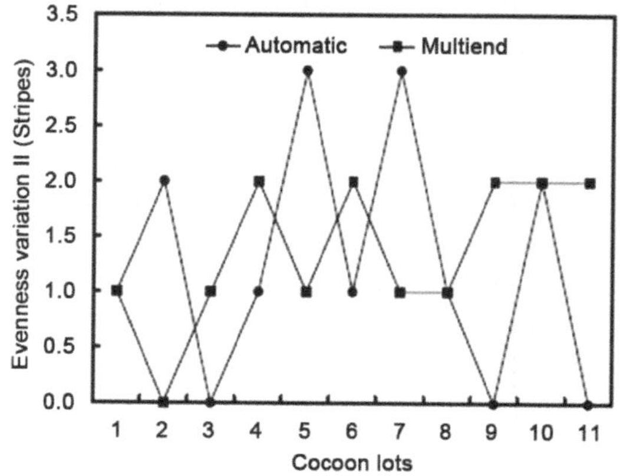

Figure 23.9 Comparison of evenness variation II of raw silk [10]

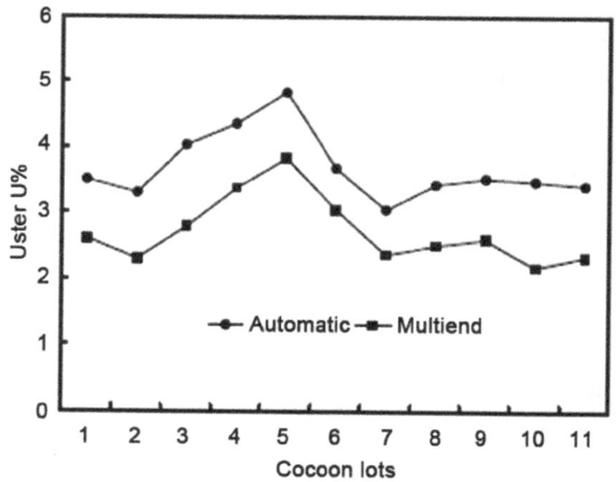

Figure 23.10 Comparison Uster U% of raw silk [10]

References

1. Shimazaki A, 1961. Studies on the statistical control of the raw silk production process – I – On the process control in reeling with fixed number of cocoons, Bulletin of Sericultural Experimental Statistics, 16, 403.
2. Yamamoto T, 1991. Properties of the silkworm race 'Ariake' with thick denier filament and the silkworm race 'Shin-Akebono' with thin denier filament, Sericulture science and technology, 30, 43.
3. Yamamoto T, Mase K, Nagasaka K, Okada E, Itsubo T, Miyajima T, Enokijima M, Kumai T, and Iszumi S, 1999. Breeding of the silkworm race, Hakugin with super thin filament, Journal of Sericultural Science, Japan, 68, 125.

4. Miura M, 1977. Stochaistic analysis of the raw silk reeling process, Applied Probability, 14, 548.
5. Miura M, 1977. Studies on the relation between the property of cocoon and quality of raw silk – II - , Journal of Sericultural Science, Japan, 46, 93.
6. Miura M, 1977. Studies on the relation between the property of cocoon and quality of raw silk – III – Stochaistic model of the variation of the size of raw silk, Journal of Sericultural Science, Japan, 46, 98.
7. Miyajima T, Yamamoto T, Mase K, Iizuka T, Nozaki M, and Kiuchi M, 2001. Induction of trimoulting larvae by the imidazole compound, triflumizole, in the silkworm races with a thin cocoon filament, Hakugin and Honobono, and the resulted cocoon characters, Journal of Sericultural Science, Japan, 70, 32.
8. Shimazaki A, 1961. Studies on the statistical control of the raw silk production process – II – On the foundation of control techniques of the size control system in the automatic reeling process, Bulletin of Sericultural export statistics, 16, 531.
9. Bai L, 2002. Analysis of the intervals of feeding ends under fixed size reeling of raw silk, Journal of Insect Biotechnology and Sericology, 71, 91.
10. Vasumathi BV, Takabayashi C, Nakajima K, and Somashekar TH, 2004. Studies on the relationship between cocoon parameters and raw silk uniformity characteristics, Journal of Insect Biotechnology and Sericology, 47.

24

Technical aspects of spider silk and applications

Summary : Compared to a number of natural and synthetic fibres, the dragline silk exhibits excellent physical and chemical properties, and enable it to withstand the adverse and extreme conditions. Despite the fact that the chemical synthesis of spider silk appears ineffective in many ways, the recombination method of producing the spider silk using biological hosts seems feasible for its production commercially. Even though a number of attempts have proved effective in the production of dragline silk at experimental levels, controlling the molecular conformation and their aggregation during the spinning for achieving properties similar to the native fibre poses problems for the future research.

24.1 Introduction

The word 'silk' covers a wide range of continuous filaments spun by the many species of Lepidoptera and Arthropoda, used for building structures for different purposes including prey capture. Spiders and silk worms produce silk filament that have superior properties compared to other silk-producing insects and more than 2500 orb weaving species are known to exist in nature [1]. Spiders have six to eight sets of glands, each producing different fiber; and these glands remained undifferentiated, early in the evolution [2]. The spinnerets, microscopic tubes originating from glands, are classified into major, minor ampullate and the term "ampulla" is used to describe the distal part of the secretary zone [3]. Unlike synthetic polymers, the biopolymers are composed of numerous monomers arranged in a strictly controlled manner [4]. A number of persistent efforts have been taken earlier to harvest and convert spider silk into fabric [5].

24.2 Spider web and spider silk varieties

The main focus has been on the construction and design of spider silk webs, before the exploring their structure and properties [6]. A number of varieties of spider webs are existent of which the orb web is the most common type. Various spider species such as Araneus and Nephila build orb web and other spider families construct tangle and sheet webs [7]. Orb-web spiders invest little energy in searching the prey and majority in silk synthesis and

DOI: 10.1201/9781003735427-24

construction of the webs. Figure 24.1 shows various threads and the web constructed by orb weaving spiders Araneus and Nephila.

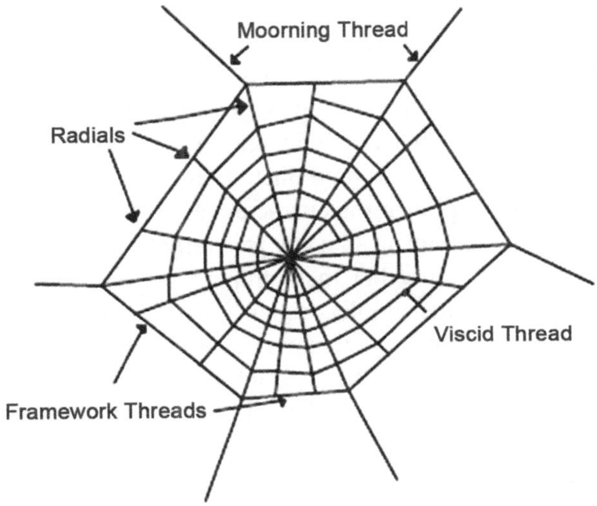

Figure 24.1 The spider web [90]

An orb web comprises of many spokes laid outward from a common point, which differs with the species of spiders [8]. The orb webs are built in such a way so as to prevent web damage caused by air drag due to capture of prey [9]. In a three-dimensional web, the energy required to stop a moving insect is dissipated mainly by breaking some of the strands, while in a two-dimensional orb web, it is achieved through stretching the spiral threads [10]. Due to high energy requirement in protein synthesis, only the damaged parts of the web are reconstructed instead of complete web and large portions of the web are repaired through the enzyme digestion and recycling [11]. The design of orb webs varies among the species, individuals and day to day for an individual and also within the web. Webs of young spiders are more circular than that of adults and hungry spiders construct smaller webs. Based on the vibration of the strands, the spider locates the prey accurately. In a web placing of capture threads in the upper half region becomes difficult since it necessitates lifting of abdomen and the spinnerets above the head and so the lower web region is of greater prey capture value and the spider increases this section of the web and decreases the extent of upper web region [12]. This asymmetry differs with factors such as age of species, weight of the spiders and web size. However the upper half is necessary to maintain the functional and stable webs.

Seven types of silk (including dragline) can be synthesized by the orb-weaving spiders by drawing liquid proteins from separate gland-spinneret complex [13]. The major portions of the orb web are the Dragline, minor ampullate and viscid silks and the dragline silk in the form of mooring threads, framework and pretensioned radial threads, which prevail over the web structure [9]. Multi-strands of spider silk are laid in the web to withstand adverse conditions and impact created by fast-moving prey. The perfume-coated dragline helps to find their mates, swing from place to place, store the food and egg and for reproduction also [14]. Minor ampullate silk produced by medium spinneret is characterized by higher elasticity and low strength [15]. Minor ampullate silks are generally finer than manor ampullate silks with better uniformity but their properties are highly uncertain [16]. Capture threads with glue droplets ranging from 7 to 29 microns are used for the aerodynamic damping of impact caused by the flying insects [17]. Capture threads produced by the flagelliform glands of Nephila Clavipes are highly compliant. Both Araneus and Nephila coat their capture threads with an aqueous solution that forms sticky droplets, which enhance the damping and harvest water from the air [18]. Its principal function is to absorb and dissipate the kinetic energy of captured flying insects. The water content of the coated substance plasticize the core threads and renders them more elastic to maintain overall toughness even after contraction through kinetically free molecular chains. The capture silk of Araneus consists of a pair of fibers coated with composition that contains 80% water and amino acids, glycoprotein and amino acids, lipids, salts and lower molecular weight compounds to the extent of 20–30%, and form regularly spaced droplets along the filament [19]. The chemical composition of the aqueous solution of the adhesive solution varies among the individuals both qualitatively and quantitatively mainly due to physical environment, diet, web recycling, and onto genetic changes in the web chemistry. While stretching, the core filament inside the droplet extends to a level of 200%, which prevents the rebound of the proirey after catching [20]. The silk fibroin also shows properties identical to major ampulate silk, when it is treated to remove the water [21]. Less than 5% crystallites (by volume) are present in capture silk, which has a high elongation and low initial modulus 3 MPa [22].

24.3 Spider silk spinning

24.3.1 Spiders and natural spinning process

Spiders are generally nocturnal with colors like orange, brown, grey and black, which reduces their visibility during daytime. The spider Nephila Clavipes measures about 2.5, with color of orange and black weighing 90–1456 mg [23]. The crursal gland and cuticular secretions are related to the silk-secreting systems of spiders and insects, which are homologous. Systematic studies of natural spinning process of spider silk have shown a marvelous process of filament making from the delicate glands at a very low temperature, using water as solvents [24]. Cephalothoraxes of the spider attached to an unsegmented abdomen, which has spinnerets at the posterior end. N. Clavipes spider has three pairs of spinnerets, namely, anterior lateral, posterior lateral, and posterior median. The largest major ampullate gland secretes dragline silk protein, exits from the anterior lateral spinneret (Fig. 24.2).

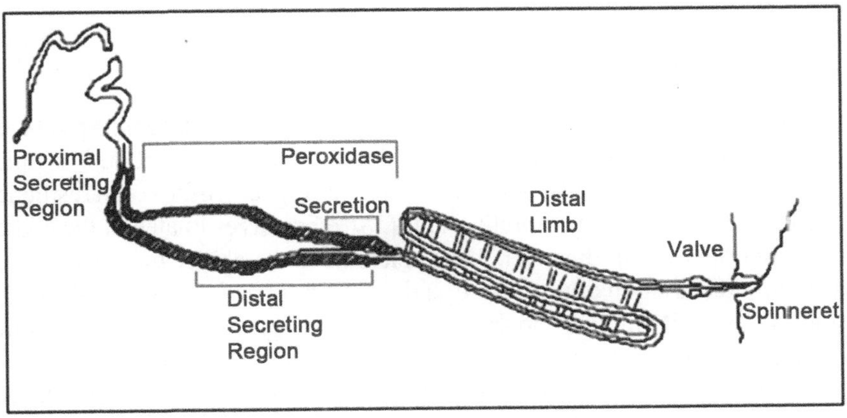

Figure 24.2 The spinning system in dragline silk [90]

Secretions of proximal region and the distal region together form spider silk. Proximal region secretions are rich in tyrosine residues, sulfhydryl linkages and acidophilic nature. They form core of the silk while secretions of the distal zone form coating of the fiber, which lacks tyrosine, sulfur contents. The duct connecting the spinneret and the distal zone causes the orientation of the protein molecules as well as water removal. Final coating consisting of four layers is given to the threads in the third limb of the 'S' shaped duct, close to the start of the drawn down taper. The duct has a thin cuticle which acts as a dialysis membrane, removes water and sodium ions out of the lumen and adds potassium, possibly surfactants, lubricants and phase separation promoters into the lumen [25]. A liquid layer is given in the final limb through an acidic bath. The valve present just prior to the spinneret controls breaking and restarting of the spinning process. The hyperbolic shape of the spinning duct imparts stress and orientation to the dope during extrusion, leads to stress induced crystallization of the material and the friction with the wall of the spider's spinning duct provides force needed to trigger the change in confirmation. Proton pumps secrete hydrogen ions to render the solution more acidic near the exit from the duct and the resulting gradient in pH, polymer concentration and potassium concentration are believed to contribute to the structural transitions. The extension of molecules under stress enables them to join together with hydrogen bonds, like zip fasteners and give anti-parallel β confirmation in the final thread. With the crystallization and aggregation of silk protein molecules, the hydrophilic groups get concealed into the interior and are changed into hydrophobic surface which enable the loss of water from the extruded thread surface [26]. The silk reabsorbs the water from the dope and thereby avoids excess loss of water.

24.3.2 Process conditions and their effects

During building, dragline silk fiber at approximately 1 cm/second is produced by a mature Nephila and during rapid descent increases to about 10 times faster rate [27]. Spider silk spun under water has higher stiffness and resilience in comparison that spun naturally in air [28]. The valve situated at the end of duct controls the silk diameter. The spiders have the ability to withstand the temperature variation of up to 30°C and humidity variation of up to 70%. The spiders can easily modify the spinning conditions by moving speed, building the webs in different times in a day, i.e. at different temperatures. Spinning speed has less influence on the diameter of the filament compared to the temperature but strong influence on the toughness. As the feedstock flows, the droplets coalesce and progressively stretched out by elongational flow to form silk filaments with canaliculi. Variability in the silk spun by the spider exists at different levels like inter-specific (between species), intra-specific (within same species) and intra-individual levels [29]. Differences exist in the composition and properties at inter-specific level and properties of the silk in the case of the intra-specific level. The factors that affect variations in silk structure and properties include body dimensions, body weight, rate and temperature of reeling spinning direction. Composition of silks produced by herbivorous spiders is rich in Glycine, Alanine and Serine and can be predicted to a larger extent. However, the silk produced by predatory spider cannot be predicted due to the different types of prey [30]. Dietary compositions of herbivorous spiders are energy rich and poor in protein content whereas the diet of predatory spiders is more diverse and rich in protein. Competition for limited or fluctuating supplies of amino acid perhaps has resulted in the evolution of two different kinds of glands to secrete protein glues and silk fibroin. The spider produces the thread on a very strict energy budget using liquid crystalline polymer. Consistent intrinsic tensile properties are exhibited in samples collected from the undistributed climbing of Argiope spider, and are not related to the weight and age of spider, relaxation of spinneret dimensions and the speed. In the vertical spinning, the spider needs to be protected from falling, regardless of its weight and this can be achieved by increased filament diameter. In the horizontal spinning, high level of spidroin and metabolic activity is unnecessary owing to lower risk level related in these directions.

24.3.3 Spider silk protein rheology

Silk protein solutions exhibit the tendency to get gel even at rest and the liquid crystalline dope differs from isotropic solutions [31]. The silk dope contains about 30–40% of a concentrated protein. Dope from the MA gland has a viscosity of about 3500 Pa second [32]. The liquid crystalline viscosity is influenced by shear rate and stress along and across the flow lines of the liquid, and is even more influenced by the concentration of the solution. By controlling solvent saturation, i.e. water content of liquid crystalline spinning solution, the spider can reduce the metabolic cost of silk production. The protein droplets obtained from the major ampullate glands of Nephila clavipes shows alkalinity and optical isotropicity indicating the absence of any ordered structures but exhibits nematic structure on loss of water. The pressure required to push the dope is reduced due by shear-induced transition to a liquid crystalline phase, localized slip of polymer solution on the wall or lubrication/ watery layer surrounding the fibroin. The elongation flow results in the strain hardening, and due to strain stretching of entangled spidroin macromolecules inhibit the capillary breaks of spun threads are limited. The effective binding potential between fibroin molecules are due to protein–protein interactions due to hydrophobic effects between non-polar amino acids, dispersion forces and ultimately the electrostatic attraction between charge-bearing arginine residues [33]. As the LCP feedstock passes down, loss of water occurs, a nematic structure is formed along with the molecular orientation. The structure follows a nematic discotic type and secondary structure from helix to sheet forms of the crystal structure. This transformation to liquid crystalline phase makes solutions of silk proteins spinnable despite their high concentration in the gland. The dilute solutions of the spider silk exhibit a shear thinning behavior with increasing shear rate and the concentrated solutions exhibit a shear thinning and shear thickening behavior. Very low critical shear rates are required to produce shear thickening behavior with the increasing concentration at pH 6.3. Interestingly, this helps the spiders to construct the webs at a low shear rate and high concentration levels. Addition of potassium ions to the dope solution in the distal limb and application of shear inside the duct result in the phase separation, formation of nanofibrils and aggregation into insoluble form. The production of filaments is enabled by the liquid crystalline behavior of spider silk in a number of ways, such as virtual freedom from the uncontrolled reorientation of molecules, less force required to spin a thread, and alignment of molecules in the dope reduces the formation of defects.

24.3.4 Structure and properties

Owing to its superior properties over existing fibrous materials, spider silk has attracted the attention from all the sections of engineering. Chemical, physical properties and their relationship of silkworm silk, spider silk have been reviewed [34]. Despite long storage, the structural features of dragline silks remain unaffected.

24.3.5 Chemical composition

Exhaustive investigations have been done on chemical composition with regard to fibroin, molecule, amino acid compositions, amino acid residues and various end groups [35]. The dragline silk protein in the case of Nephila clavipes has molecular weight 720,000 and in the case of proteins obtained from glands and spun fibres it is 96000 to 520000. Major ampullate silk fibroin of dragline spider contains multiple repeats of long polyamide block and glycine/proline-rich blocks [36]. About 5–10 residues of polyalanine form a beta sheet and account for most of the crystalline fraction. Dragline silk can be considered as the block copolymer with polyalanine blocks and of glycine-rich domains that imparts stiffness, strength and toughness. The chemical compositions of various silks vary with the type of function they need to perform and fibroin composed of twenty odd alpha amino acids. Fibroin content of the Nephila has been found to be about 90% at 65% RH. All types of silk share four types of amino acid motifs comprising of GPGGX/GPGQQ, GGX, Poly(Ala) or Poly(Gly-Ala) and spacer sequences to prevent premature fiber formation within glands, align various structural regions within the fiber and to provide additional surfaces for critical interactions. GPGXX pentapeptide conforms to a spiral and the proline residue behaves as the focal point for the retraction energy after stretching. The small side chains in the polymer consist of glycine, alanine and serine and the polar substance consists of aspartic acid, threonine, serine, glutamic acid, tyrosine, lysine, hystidine and arginine. The various amino acid contents are found to be Alanine – 29%, Glycine – 40%, Proline – 3%, Tyrosine – 4%, Glutamine – 10%. Hydrogen bonds between carbonyl and amide groups, Vander vaal interactions in the short chain amino acids are found between the molecules. Glutamine, glycine residues in the amorphous regions contribute to the hydrophilic nature of the spider silk. Table 24.2 shows amino acid contents of dragline silk and comparison with other protein fibres like wool keratin, silkworm silk [37]. In addition to the fibroin, other classes like glycoprotein, inorganic salts, sulphur-containing compounds, amino acids, and ionic forms of amines are also present in the spider silk. These chemicals perform important functions such as identification of species, regulation of water content of the web, protection against microorganisms. The antimicrobial properties of the spider silk are due to the presence of 12-methyltetradecanic acid and 14-methylhexadeconic acid to the minor amounts. There are wax-like esters on the surface of the spider silk.

24.3.6 Action of solvents, supercontraction

Besides chymotrypsin, spider silk has resistance of many of the common solvents and enzymes. The protein of spider silk dissolves partly in mineral acids and alkaline hydroxide and dissolves completely in concentrated solutions of LiBr, Lithium thiocyanate or $CaCl_2$, ethyl alcohol, mercapto ethanol, and mixture of boiling HCl and propionic acid [38]. Under the presence of urea, unrestrained ampullate fibers undergo supercontraction up to 60–80%, which also increases with the polarity of the solvent e.g.CH_3OH, H_2O and leads to decrease of modulus values, breaking strength [39]. This type of contraction is not observed in minor ampullate silk, viscid silk to this extent. Supercontraction of dragline silk also takes place in air at about an RH% of 90% or more and takes up slackness of the webs, restores shape and tension after prey capture [40]. The super contraction ratios have been found to be 0.55, 0.82 for Araneidae and Nephilengys, respectively [41]. Super contracted fibers can be re-extended to their original length in the same medium or in water. Contraction leads to changes in conformations and orientation of the molecular chains due to rupture of hydrogen bonds [42], and the deformation energy is stored in the form of conformational entropy. Initial exposure of spider dragline to moisture leads to development of supercontraction stresses, which reaches initially the yield stress and increases up to 22% of breaking stress depends on the effect of moisture on chain mobility, the rate of moisture enabled conformational changes, and the rate of stress relaxation of super contraction [43]. Supercontraction affects the morphological parameters as well as the physical properties of the dragline silk [44]. In Nephila substantial fraction of the glycine glutamine, tyrosine, serine and leucine exhibit increase in molecular motions than polyalanine beta sheet regions. Supercontraction indicates an increase in the random coil conformation with increase in elasticity, decrease in strength. The X-ray diffraction angle of the

(200), (120), (210) reflections is not altered by supercontraction but long period is reduced from 8.4 to 5.8 nm and crystalline orientation decreases. Investigations have been carried out on the influence of supercontraction over diameter, birefringence, and force elongation relationship for different species of spider silk and are mainly due to the plasticizing effect of the solvents. The birefringence of the core remains unaltered in the case of Nephila even after contraction whereas that of the skin decreases reduced to half, i.e. to 0.016.

24.3.7 Dragline silk: Macroscopic structure

Dragline spider silk has golden yellow colour and a circular cross section having 7 μm average diameter. The fiber is not related to sericin or glue-like protein, similar to that of silkworm silk. Studies of the dragline silk structure show a skin core structure with a weak skin [45], possibly caused by the two different secreting regions responsible for the production of the silk filaments. The diameter of the silk produced by the spider is related to the weight of the spider to a small extent. Dragline silk consists of twin filaments with circular cross section, stuck together and aligned parallel to the filament axis [46]. Significant variation of the fiber diameter exists, along the fibre length because of the way in which the silk filament is produced and laid, to the extent of ±20% of the mean value [47]. The density of the spider silk is about 1.33–1.35 g/cc [48], which is similar to that of other protein fibres like wool, silkworm silk but lower density values have also been reported in the literature. The weight of female Nephila increases exponentially from juvenile to mature in autumn and the density also varies during these periods from 1.271–1.289 g/cc. Outer layers of Nephila madagascareinsis show a thin peripheral layer with elongated cavities like caniculus, oriented parallel to silk fiber axis. These domains act as stress concentration points, crack initiators that inhibit the spreading of the cracks by delaying rupture of the fiber by taking up the energy [49]. The inhibition influence of these areas is due to self-repairing of cracks by solidification under high stress areas, plasticizing influence by hydration to act as fluid shock absorbers, and act as a lubricant to reduce the interfibrillar friction.

24.3.8 Fine structure of dragline spider silk

The dragline spider silk comprises of semi-crystalline polymeric structure with many small crystallites between amorphous regions [50]. The areas rich in alanine (40%) are situated in the crystalline domains of the silk fibre having beta sheet structure. Around 60% of alanine is found in the weakly oriented and unaggregated beta sheets. In the case of dragline, a less ordered secondary structure is found in the glycine rich chain segments, and helices can occur locally, stabilized by hydrogen bond. Glycine residues are partially found in beta sheets but contribute to helical structures showing 3-fold symmetry. The presence of bulky glutamine residues limit the growth of beta sheets and force the formation of loops, tie chains and the surrounding amorphous matrix. Poly(Gly-Ala) region can form a similar structure with lesser interactions, leading to lower tensile strength, which happens in minor ampullate silk. Disordered regions of major ampullate silk in the relaxed fibers, follows the helical conformation under tension. Under tension, the glycine rich sequences are able to form large-scale beta sheet crystal structures that are less perfectly ordered than the polyalanine crystals. The composition of repeat units varies to the extent of few nanometers, resulting variations in intersheet spacing and order and is termed non-periodic lattice crystals (NPL). Polyalanine crystals are surrounded by a non-periodic crystal lattice' formed in the glycine-rich sequences. The crystal state of dragline silk assumes various unit cell dimensions in their water-soluble and -insoluble states after spinning, despite their being orthorhombic structures. The change from soluble to insoluble state involves decrease in the distance between the overlaying sheets to an extent of 18.3%. Beta sheet conformations of the protein macromolecules lead to the formation of crystallites of at least 2×5×7 nm determined from the (002), (200), and (120) peaks. The diffraction pattern obtained from a single silk of Nephila clavipes yields reflections, showing orthogonal unit cell of a = 0.93 nm, b = 1.04 nm, and c = 0.70 nm. SAXS shows an axial period of ~79A°, and the smallest ordered regions exist in the range of 20A° across. TEM results show higher domain size of the crystallites, i.e. 70–500 nm including non-periodic lattice regions. The degree of crystallinity is about 30%, which exists in both the oriented state and partially oriented state with the orientation function of 0.776 – 0.979. The variation of reeling speed of silk by spiders apparently does not influence the lateral packing of the molecular chains and the size of the crystallites; however, it alters the azimuthal width of the crystallites. Post-spinning drawing of silk obtained from 'forced silking' process results in the partial degradation of crystalline region with improved amorphous orientation [51]. In general the fine structure of a fiber may be described using one phase model with continuous defects in a regular-oriented system or two-phase model with separate crystalline and amorphous regions. However, in many occasions, these models become inadequate description for explaining the properties of the polymers, fibers and composites (Figure 24.3), which in turn require an additional phase "interphase", a morphologically distinct region, which is also known as intermediate phase or oriented amorphous phase [52]. This intermediate phase has physical properties that are different from both amorphous and crystalline phases. Formation of interphase takes place in a polymer when the orientation of the crystallites are not convenient for the reentry of the chains and when the crystallites are small the formations of inter phase increases, e.g. dragline silk. The crystalline fraction of dragline silk ranges between 0.1 and 0.4, and does not help the mechanical properties and necessitates an additional interphase, ranging between 1 nm and 4 nm around every crystallite. The interphase is between 1.145 g/cc and 1.157 g/cc, which differs from the amorphous state (1.139 g/cc). Moreover, the glass transition temperature (T_g) of the interphase is greater than the amorphous phase having the modulus about 6 times greater than the amorphous state.

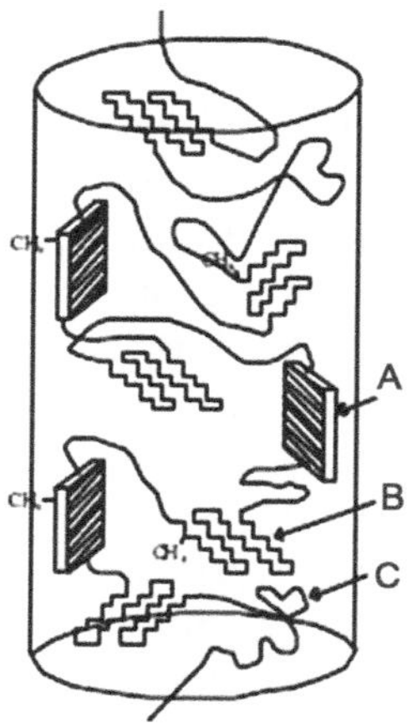

Figure 24.3 Secondary structure of dragline silk [90]
A – Crystalline region; B – Oriented amorphous region; C – Amorpous region

24.3.9 Optical properties

During the silk protein extruxion process, the optically isotropic protein secretions of MA glands are changed into anisotropic liquid crystals and perfect three dimensional crystallites. The liquid protein of Nephila Clavipes liquid has a low value of birefringence ranging between 0.01 to 0.02.The form birefringence of the fibres contributes more significantly to the total birefringence. Owing to the attractive pattern and light reflecting properties of the web, insects are attracted towards it [53]. The silvery caraphase of the Nephila helps to reflect sunlight, which reduces the overheating of insects. Light reflection properties of the spiders, webs contribute to maintain the physiological conditions and prey capture. Spiders and webs are capable of protecting themselves from adverse conditions of overheating due to sunlight [54]. Forced "evaporative cooling" in the body also takes place above 35°C.

Definite dark and bright patterns existing over the body of the spiders also help them to camouflage themselves from other insects and predators [55]. Tiny scales on the bodies of the matured male jumping spiders reflect UV light, to identify and communicate with other spiders [56]. Due to their ultra violet reflecting properties, the brightly coloured Nephila spiders with both bright and dark markings are able to attract more insects. The webs having the reflection patterns like many flowers in UV light, attract and trap the insects more than unreflected portions [57].

24.3.10 Mechanical properties

The factors like composition of amino acids, insect size, diet conditions, body temperature and drawing speed strongly influence the mechanical properties of the dragline silk [58]. The frame thread of dragline silk should be strong enough to be an energy absorber so as to catch the insects and enable safe escape of the spiders. A controlled environment for conducting different tensile tests is required owing to the sensitivity of the spider silk to humid conditions. With increasing relative humidity, the stress developed in the fibres increases very significantly with decrease in yield stress. Stiffness, strength and toughness of the dragline silk increase with the strain rate, which is an essential criterion in the prey capture or during the fall of the spider. The stiff radial threads have time-dependent visco-elastic properties while the pliant capture threads have time-independent elasticity. Stress relaxation in the silk depends, largely, on the initial stress level applied to the fibre and upto 45% decay in stress levels are obtained if the fibres are allowed to relax within the yield stress. In the orb webs, the longitudinal waves are transmitted with little or no attenuation and resonance effects are not found in the frequency range at which the preys are expected to produce the impact [59]. Energy dissipatio through stretching and relaxation of the web is too weak; however, the main importance of the web lies in the dissipation of the kinetic energy of the insect upon impact on the web. Similar to high performance fibres, the molecular deformation pattern of Nephila edulis reveals a linear relationship in shifting of wave number in Raman Spectroscopy [60]. Structural analysis of spider dragline silk carried out through atomic force microscopy reveals a fibrillar structure with pleated fibrils at the core of the fiber, which can be unfolded with the extension [61]. Spider silk displays a force extension curve that rises to a large force quickly and maintaining it over large extensions [62]. Stretched fibers usually snap back because the surface crack propagates only upto the point in the crystallites. Measurement of modulus values have been found to be difficult on account of wide variation in diameter of fibres and the initial modulus of Nephila silk is close to 12.70 GPa with an elastic strain about 3.0%. The breaking strength of silk increases linearly with increasing spider weight [63] and breaks at a stress about six times the spider's weight. The stress–strain curve of dragline enters non-linear region when the stress exceeds twice the spider's weight. The average tensile strength of the dragline of Nephila clavipes is almost three times that of Bombyx mori (1.3 and 0.5 GPa, respectively) with the elongation at break of 40% and 15%, respectively. The highest strength of dragline silk has been measured for Nephila clavipes (1.7–2.9 GPa) and lowest for Araneus diadematus (0.8 GPa) [64]. Tensile strength of spider silk reduces, when it is subjected to acid rains, UV radiation. Spider silk exhibits a high level of torsional resistance, with a shear rigidity of 2.38 GPa, which is higher than all textile fibers. Higher torsional stability is essential in the case of dragline to serve as the lifeline for the spiders. Spider silks can undergo large tensile and compressional deformations (20–40%) without any evidence of fracture or kink formation [65]. The compressive modulus has been found to be around 0.58 GPa. The ability of spider silk to resist transverse compression is lower than many of the textile fibers like Kevlar 29, nylon 5, polyster and wool. There is a great difference with regard to breaking energy between both the silks (16×10^4 and 6×10^4 J/kg), respectively. There are many kinds of fracture behavior in the spider silk, like brittle failure at one place, such as happens in ceramics, multiple brittle failure, and failure with elongation. Ductile failures at low strain rates can also be seen in the dragline of Nephila clavipes.

24.3.11 Thermal and electrical properties

The thermal properties of spider silk are rendered more interesting by a very low second order transition, high degradation temperatures and sustained properties at extreme temperatures [66]. The thermo-mechanical properties of the silk can be predicted by means of its chemcial composition and arrangement of the structural elements [90]. The glass transition temperature of the spider silk lies in the range of −50°C to −60°C and the transitions observed at −20°C and 30°C remain still unexplained. Thermal expansion coefficient (a) value of dragline silk has been found to be $-10.6 \times 10^{-4}C^{-1}$, which reduces further with the increasing draw rates. Molelcular organization of Araneus diadematus silk fibroin is unaffected by exposure to 100°C for over 50 h [67] and the functional properties are retained upto 180°C without affecting the initial modulus. Thermal decomposition occurs above 250°C. Many natural biopolymers, like wood, wool, silk fibroin and collagen display piezo-electric effect. The dipole orientation in polarized species containing carbon, oxygen, nitrogen and hydrogen, leads to piezo-electricity. Spider dragline has lowest piezo-electricity compared to other natural polymers. The piezo-electric is not seen in the randomly distributed dipoles.

24.4 Regenerted spider silk

24.4.1 Forced silking (reeling) of spider silk

In order to reel about 3–5 mg of silk at a time, reeling devices have been developed to force silking of dragline from the glands of anaesthitised Nephila clavipes. In forced silking under anesthetised conditions, the cross section of the filament changes from circualar to ellipsoid with variation in diameter along the length. However, forced silking of fibers under unanesthitised state leads to properties identical to that of normally spun spider silk.

24.4.2 Spider silk protein: Artificial route

Due to its superior properties, the protein of dragline silk has gained scientific interest. The aspects relating to structure, genetic engineering and expression of silk proteins of silkworm and spider as well have been evaluated [68, 69]. Earlier, attempts have been made to splice silk genes into two different cell lines with bovine mammary cells and hamster kidney cells for producing great quantity of recombinant proteins. The genetical sequencing of the flagelliform silk of tropical spider Nephila clavipes and N.Madagascareinsis has been effectively done. A method has been developed to isolate cDNA, capable of expressing spider silk proteins, and for producing recombinant protein [70]. Recombinant DNA technology for microbial proteins [71] appears to be advantageous as compared to that of chemical systhesis due to low cost, rapid preparation and absence of byproducts. In spite of numerous steps involved in chemical synthesis, the yield and molecular weight seems to be very low with high polydispersity, whereas biosynthesis of polypeptide molecular weight to the extent of 40,000 with monodispersity could be achieved. Various recombination techniques for protein synthesis have been successfully reported using microbial sources [72], transgenic techniques involving mammary cells [73], plant cells and leaves [74]. Table 24.3 summarizes the various attempts in recombinant synthesis of spider silk protein. From the known sequence of spider silk protein, genes are constructed and expressed using E.Coli as the host, which has successfully used earlier for silkworm silk also [75]. E.Coli as the host yields the protein to the extent of 10 mg/g in the wet state [76] and 1–2 mg/g of E.Coli after purification, which is higher than that obtained using Bascillus subtilis [77]. Genes of spider dragline silk have been inserted into mammary gland cells along with regulatory elements. Insertion of the genes into single cell goat eggs has been tried to produce water-soluble silk protein [78]. Few milligrams of genetically engineered silk-like protein have been successfully produced based on the sequence of spider silk protein. Recombinant proteins of Nephila clavipes using gene synthesis route have also been attempted in endoplasmic reticulum of tobacco leaves, and potato leaves and tubers. Such proteins show excellent heat stability while purification and is relatively cheaper to produce than bacterial method. Producing recombinant silk proteins is relatively easier than regeneration, spinning and drawing which are the bottlenecks in the whole process sequence. The protein sequence of Nephila clavipes, synthesized by cloning shows high susceptibility to enzymatic cleavage near the inverted repeats. The recombinant protein afer purification rapidly self-assembles, resulting in insoluble microfibrils and become a constraint in the subsequent conversion into similar fibres. In order to control the conformational changes through prevention of hydrophobic interactions between the overlaying sheets, self "sterical-triggers" have been introduced into the system [79].

24.4.3 Regeneration or spinning

A number of efforts have been directed towards regeneration of silk proteins into fibres [80]. Attempts have been made for solution spinning of silk protein in different solvent–nonsolvent systems such as hexa-fluoro isopropanol, HFIP isopropanol, HFIP methanol, guanidine hydrochloric mixture followed by drying at room temperature using special spinning set up [81]. Efforts have been taken to produce dragline by means of a biometric rig that mimics the biological set up and electro spinning technique method with a blend of PVA and silk protein (50/50 by weight) at 35 kV. Etched silicon wafers with the aperture sizes of 80–160 μm and syringe have been successfully used to extrude the filaments from protein solutions of dragline silk with a diameter of 20–80 μ and surface characteristics similar to the native filaments [82]. The wider variation in diameter is attributed to the uncontrolled reeling of the filament in plastic condition. Blunt point hypodermic stainless steel needles with a diameter of 150, 175 and 250 μm attached to disposable syringes have also been used [83] as spinnerets for producing silk fibres. The filaments are subsequently drawn in air at 8 cm/s under the wet condition followed by drying under tension and the filaments appear to withstand a total draw ratio of 9 in two steps without rupture. Microscopic analysis of freshly regenerated fibres exhibits low density values on account of voids crated due to the transfer of solvent, which remains even after the second drawing, while spongy appearance is modified into the fibrillar structure. The highest strength measured in the regenerated and drawn dragline silk fiber remains about 320 MPa with the modulus value of about 8.0 GPa compared to 875 MPa and 10.9GPa of native fibers used for regeneration. The fraction of alanine β sheet structure is found to increase linearly with the extent of drawing applied to the specimen [84]. The two stage drawn fibers exhibit fraction of β sheet conformation as high as 65%, which is closer to the native fibers. While no preferential molecular order is observed in as-spun fibers, significant crystalline alignment with respect to the fiber axis is achieved when the spun fibers are stretched in water for the second time. Owing to the presence of unoriented crystalline regions, the drawn fibers show low average orientation compared to the native fibers. Fibers regenerated from proteins of Nephila edulis spider shows a diameter of 9 μm and rough surface along the wrinkles. An initial modulus of 6.0 GPa and breaking strength of 0.11 to 0.14 with elongation at break of 10% and 27% has been reached in this case. The values appear to be much lower than native spider silk except for the elongation at break [85]. Regenerated spider silk exhibits dichroism only after drawing, indicating the random distribution of ordered beta sheets in the unstretched filament. The regeneration of fiber silk proteins results in the conformational changes to restore the original structure after spinning. But, the reconversion occurs only to a lesser extent and achieving higher beta sheet conversion during spinning and post-spinning operation is very necessary to get better properties.

24.4.4 Uses

Despite its limited availability, the dragline silk is widely used in defense and medical applications [86]. The potential areas of applications have also been explored earlier [87]. As dragline silk compares well with high performance fibres such as kevlar with regard to structural similarity and comparable properties, it becomes more suited for applications where high performance in terms of physical properties are necessary [88]. Until World War II, spider silk was used as crossed-hairs in optical devices including microscopes, telescope and bomb-guiding systems. Silk strands of the web have an ability to elongate when an insect is caught, convert the prey's momentum, i.e. the kinetic energy into heat, and dissipate about 70% of the kinetic energy. The web also gently rebounds so as not to catapult the insect back out. This ability to dissipate energy at very high strain rates makes spider silk suitable for body armor system and unwanted one high performance applications such as bullet proof vests. A very low glass transition temperature of −50°C to −60°C enables it to absorb sudden shocks at low atmospheric temperature and makes the spider silk suitable for parachute applications. But, supercontraction in water is unsuitable in the fabrication of parachutes. Earlier use of spider silk in the form of web, rather than a fiber, includes wound dressing to help blood clot and fishing nets. Biomedical applications, such as degradable suture, has been reviewed earlier [89]. Spider silk protein can be used to coat the medical implants for better performance. Surgical thread, biomembranes and scaffolds for tissue engineering are the possible areas of application in biomedical and biomaterial fields due to low inflamatory potential of silk proteins and antithrombic nature. Recombinant spider silk has potential applications in sutures for eye surgery, artificial tendon, ligaments for knee construction. Spider silk with higher safety coefficient can be used in structural applications like elevator ropes, bridges and pillars. Spider silk can be used as reinforcement for other fibres and can be incorporated directly as it is spun from the glands of the *N.clavipes* [88]. For another use, the spider silk of Nephila madagascareinsis is burnt by repeated treatment with tetraethyl orthosilicate and permitting to dry, and then baking at 420°C. It causes shrinkage of silica to form hollow silica tube having a diameter of just 1 μm, and can be used as hollow optical fibers to carry light beams in the nanoscale optical circuits or as nanoscale test tubes. They are applicable in near field microscope [86].

References

1. Craig, Catherine L, Annual review of entomology, 42, 231–267, CAN 1997:39078.
2. Hayashi CY, Blackledge TA, and Lewis RV. Molecular biology evolution, 2004, 21(10), 1950–59.
3. Volrath F, and Knight DP. International Journal of Biological Macromolecules, 24(1999), 243.
4. Guhrsa KH, Weisshartb VK, and Grossea F. Review in Molecular Biology, 2000, 74, 121.
5. Mike T. Discover 1992(5), 32.
6. Weeins Jr. HV, Edwards Jr. GB, IFAS extension news, EENY, 229.
7. Shao Z, and Vollrath F. Polymer 1999, 40, 1799.
8. http://pa.msu.edu/~askst/071697.html, June 2005.
9. Lin LH, Edmonds DT, and Vollrath F. Nature, 1995, 373, 12.
10. Becker N, Oroudjev E, Mutz S, Cleveland JP, Hansma PK, Hayashi CY, Makarov DE, and Hansma HG, Nature materials, 2003(2), 278.
11. Ko FF, Kawabata S, Inone M, and Niwa M. http://www.materials.drexel.edu/fml/classes biomaterials/ko.pdf, 2005.
12. Herberstenn ME, and Heiling AM. Animal behavior, 1999(58), 1241.
13. David L Kaplan. Polymer degradation and stability, 1998, 52, 25.
14. Ryder ML. Textile magazine, 1999, 2, 20.
15. Hayansha SY, Shipley NH, and Lewis RV. International Journal of Biology and Macromolecules, 1999, 24, 271.
16. Work RW. Textile Research Journal, 1976, 47(10), 650.

17. Perkins S. Science news, 164, 30, August 2003, 141.
18. Volrath F. Reviews in molecular biotechnology, 2000, 74, 67.
19. Higgins LE, Townley MA, Tillinghast EK, and Rankin MA. Journal of Arachnology, 2001, 29, 82.
20. Hayashha CY, and Lewis RV. Bioassays 2001, 23(8), 750.
21. Gosline J, Guerette P, Ortlepp C, and Savage K. Abstracts compendium of biochemistry and physiology, Part B, 2000, 126, S42.
22. Gurett PA, Ginzinger DG, Weber BHF, and Gosline JM, Science 1996, 272, 112.
23. Garrodo MA, Elices M, Viney C, and Riguiero JP. Polymer, 2002, 43, 1537.
24. Service RF, Science 2002, 295, 419.
25. www.smithinst.ac.uk/projects ESGI46/ESGI46-Spider silk report/Report /silk.pdf.
26. Ferguson J, Hudson NE, Warreb BCP, and Tomatarian A. Nature, 1987, 325, 234.
27. Guess KB, and Christopher Viney. Thermochimca Acta 1998, 315, 61.
28. Liu Y, Shao Z, and Vollrath F. Chemical communication 2005(5), 2489.
29. Garrido MA, Elaces M, Viney C, and Reguiero JP. Polymer 2002, 43, 4495.
30. Craig CL, Hsu M, Kaplan D, and Pierce NE. International Journal of Biology and Macromolecules, 1999, 24, 109.
31. Hudson S. Protein based materials, Burkhauser, Boston, 1997, 315.
32. Koji N, Bico J, Clasen C, and McKinley GH. web.rnit.edu/nnf/publications/GHM71.pdf.
33. Braun FN, and Viney C. arxiv. Org/abs/physics/0210044.
34. Kaplan DL, Lambardi SJ, Muller WS, and Fossey SA. Biomaterials, 1991, 3–53. CAN 120:25648b.
35. Holland GP, Lewis RV, and Yarger JL. Journal of American Chemical Society. 2004(126), 5867.
36. Riekel C. Nuclear methods and instruments in physics research bulletin, 2003(199), 106.
37. Kaplan DL, Mello CM, Arcidiocono S, Fossey S, Senekal K, Muller W. Protein based materials, Birkhauser, Boston, 1997, 104.
38. Yang, Zhitong, Luvak, Oskar Seidal, Andreas, La Verde, George, Zax, David B, and Jelinski Lynn W. Journal of the American Chemical Society, 2000, 122(37), 9019.
39. Winker S, and Kaplan DL. Reviews in molecular biotechnology, 2000, 74, 35–93.
40. Weisshart K. Reviews in molecular biotechnology, 2002, 74, 5–66.
41. Fossey SA, and Tripathy S. International Journal of Biology and Macromolecules, 1999(24), 119–125.
42. Shao Z, Vollrath F, Sirichaisit J, and Young FR. Polymer, 1999, 40, 2493.
43. Riguero JP, Elices M, and Guinea GV. Polymer, 2003, 44, 3733.

44. Yang Z, Liivak O, Seidal A, Laverde G, Zax DB, and Jelinski LW. Journal of American Chemical Society, 2000(122) 9019, CAN, 133: 346091.
45. Madsen B, and Vollrath F. Naturwissenschasften, 2000, 87, 148.
46. Osaki S. International Journal of Biological Macromolecules, 1999, 24, 283.
47. Dunaway DL, Thiel EL, Srinivasan SG, Viney C. Journal of Material Science, 1995, 30(16), 4161.
48. Sirichaisit J, Young RJ, and Volrath F. Polymer, 2000, 41, 1223.
49. Shao Z, Hu XW, Frische S, and Volrath F. Polymer 1999, 40, 4709.
50. Yves T. Macromolecules 1994, 27(25), 7378.
51. Carmichael S, Barghout JYJ, Viney C, International journal of biology and macromolecules, 1999, 24, 219.
52. Flory PJ, Yoon DY, Nature 1978, 272, 16th March, 226.
53. Craig CL, ACS Symposium Series, 1994 544, 59, through CAN 120:73719x.
54. Maloney LT. Trends in cognitive science, 2003, No.7, 473.
55. Tso IM, Lin CW, and Yang EC. Journal of Experimental Biology, 2004, 207, 2631.
56. Soh N. The strait times interactive, 2004, 12.
57. Wickelgreen I. Science news, 135, 27th May 1989, 330.
58. Dai L, Zhang Y, and Ou Yang Z. Thin film solids, 2003, 438.
59. Naftilan SA. International Journal of Biological Macromolecules, 1999, 21, 289.
60. Young RJ, Eichhorn SJ, Sirchaisilt J, Brookes. Material research proceeding, 2002, 711, (135–146) through CAN 137:2002, 264290m.
61. Li SFY, Mcghie AJ, and Tang SL. Journal of Vacuum science and technology. A 1994, 12(4, Pt2), 1891–1894, CAN 1994, 121, 111319x.
62. Smith BL, Schaffer TE, Viani M, Thompson JB, Fredrick NA, Kindt J, Belcher J, Stucky GD, Morse DE, and Hansma PK. Nature 1999, 399, 761.
63. Osaki. Spider silk as mechanical life line, Nature 384, 5th December 1996, 416.
64. Mahoney DV, Vezie DL, Eby TK, Adams WW, and Kaplan D. ACS Sympossium, 1994, 196–210 through CAN 1994, 120, 73847n.
65. Weiss P, Science News, 2002, 161, 38.
66. Porter D, Vollrath F, and Shao Z. European Physics Journal, 2005, E16, 199.
67. Barghout JYJ, Czemuszka JT, and Viney C. Polymer, 2001, 42, 5797.
68. Gregory H, Frank Diaz, Caroline Jakuba, Tara Calubro, Rebecca L, Jingsong Chen, Helen Lu, John Richmond, and David L. Biomaterials 2003, 24, 401.
69. Stokstad E. Science 287, 24th February 2000, 1378.
70. Lewis RV, Xu M, and Michael B. US Patent 5728810, 17th March 1998.
71. Anon. Nature Biotechnology 2001, 19(6)573.
72. Heinemann K, Guhrs KH, and Weizhart K. Chemical fibres international, 2000, 50(2), 44.

73. Karatzas CN, Turner JD, and Karatzas AL. WO 9947661, 1999, 0923.
74. Guhrs SJ, Heinz K, Frank G, and Udo C. Nature biotech, 2001, 19(6), 573.
75. Sullivan DO. C&E news, 1988, July 25, 24.
76. John PT, McGrath KP, Digirolamo C, and Kaplan DL. Polymer material science and engineering, 1995(72), 80–81, CAN 123, 1995, 134313w.
77. Fahnestock, Stephen R. US Patent 6268169, 31st July 2001.
78. Martin DC, Jiang T, and Bucko CJ. Protein based materials, Birkhauser, Boston 1997, 339.
79. Winkler S, Szela S, Avtges P, Valluzzi R, Kirschner DA, and Kaplan D. International Journal of Biology and Macromolecules, 1999, 24, 165.
80. WO 2003060099, July 24, 2003.
81. Trablic KA, and Yagger P. Macromolecules, 1998, 31, 462.
82. Seidal A, Liivak O, Calve S, Adaska J, Ji G, Yang Z, Grubb D, Zax DB, and Jelinski LW. Macromolecules 2000, 33(3)775.
83. Kimberley A, Trablic, and Paul Yagger. Macromolecules, 1998, 31, 462.
84. Dong A, Xu M, Middaugh CR, and Lewis R. Polymer preprints, 1991, 31(1).
85. Shao Z, Vollrath F, Yang Y, and Thogersen HC. Macromolecules, 2003, 36, 1157.
86. Gellynck K, Verdonk P, Almqvist KF, Nimmen EV, Gheysons T, Mertens J, Langermove IV, Kiekens P, and Vetruggen A. European cells and materials, 2000(10)2, 45.
87. Penman D. New Scientist, 2003, 22, March 20.
88. Fox D. New Scientist, 1999, 24, April 38–41.
89. Kaplan DL. Polymer degradation and destabilization, 1998, 59, 25–35.
90. Saravanan D. Spider silk – Structure, properties and Spinning, Journal of Textile and Apparel, Technology and Management, 5(1), 2006, 1–20.

25

Morphology studies on electrospun nano silk fibre

Summary : Investigation on the influence of solvent concentration on the morphology of electrospun silk fibre (Bombyx mori) shows that higher concentrations help to produce superior quality silk fibres. The solvent concentration comprising of lithium bromide and calcium chloride, methanol and water in the ratio 1:2:8, has been used in the study. Increase of solvent concentration leads to fibres of higher diameter. Nano fibres produced by using the solvent lithium bromide show slightly finer diameters and better distribution, particularly at lower concentrations compared to using calcium chloride as solvent. Scanning electron microscope has been used to study diameter of fibers.

25.1 Introduction

Silk fibres of the *Bombyx mori* type have been first produced using aqueous hexafluoro-2-propanol [1, 2]. The diameter of fibre has been found to be between 6.6 nm and 200 nm. Subsequent investigations produced electrospun silk nanomats of greater porosity [3]. Solvents that are presently being used for dissolving silk compromise of lithium bromide, calcium chloride, and certain solvents like aqueous hexafluoro-2-propanol [HFIP] [4, 5]. The solvents have been so selected on considerations like ease to dissolve, safety and necessity of silk degradation. Even though aqueous-derived silk fibroin scaffolds are known to degrade faster than HFIP-derived scaffolds owing to greater surface area, HFIP is highly toxic and thus needs to be used with caution in controlled amounts. The electro spinning cannot easily produce uniform, finer and bead-free pure silk nano fibres [6]. As the electrospun solutions containing silk are of low viscosities, polyethylene oxide is used along with silk and subsequently removed after spinning [7]. But the biocompatibility and mechanical properties of silk is affected by residual polyethylene oxide [8, 9]. The yield of nanofibre can be increased by greater concentrations of regenerated silk in formic acid, together with optimum process factors [10]. It has been found that silk concentrations below 5% could not spin fibres, whereas during electro spinning, many beads and droplets have been produced during electro spinning at concentration below 10% [11]. Continuous electrospun have been produced at concentrations ranging between 12% and 19.5%, irrespective of electric field and distance. The average diameter was larger in higher concentrations than at lower concentrations. More number of

DOI: 10.1201/9781003735427-25

beads are formed at low concentrations. Coarse fibres are formed at very high concentrations owing to restriction of flow of the polymer at the needle tip [12]. Even though good deal of literature is available on the electrospinning of silk, there is no conclusive finding among researchers on the influence of solvent concentration on morphology. There are insignificant comparative studies based on two or more solvents, to examine the effect of solution concentration on morphology, under the same experimental design. The chapter highlights the production and comparison of pure electrospun silk nano fibres of *Bombyx mori* type, from two solvents with chosen concentrations of electrospinning solvents.

25.2 Technical details

Bombyx mori silk fibres have been used. The fibres have been boiled using a very small % of sodium carbonate. Degummed silk fibres have been dissolved in lithium bromide and then heated for certain duration. The lithium salts have been removed from the cooled silk solution by dialyzing against distilled water with cellulose dialysis tubing [13]. The solution so dialyzed has been filtered and subjected to evaporation at room temperature to produce dry silk fibroin. Also, a solvent system comprising of calcium chloride: methanol: water has been used to dissolve the degummed silk. It has then been dialyzed and evaporated to produce dry silk fibroin films. For electrospinning, the silk fibre films have been dissolved in formic acid for time duration, resulting in viscous polymer solutions of particular concentration, which have then subsequently filtered. This has been done for silk fibres obtained through lithium bromide and calcium chloride, separately. Electrospinning has been done using similar chosen factors for both types of silk fibres. A special type of scanning electron microscope has been used to study the morphology of the electrospun fibres. Photoshop software has been used to analyze number of fibre segments randomly to obtain a mean diameter for each nonwoven web, and examine diameter distribution.

25.3 Findings of the investigation

Figures 25.1 and 25.2 depict the SEM images of electrospun silk at recorded concentrations. The diameter distribution reveals that electrospun lithium bromide nano silk fibres are slightly finer than electrospun calcium chloride nano silk fibres [13]. There are almost no electrospun fibres below 10% concentration. But there is formation of beads. At lower concentrations, the fibres appear in entangled and open mesh-like form. More discontinuous fibres are also formed at low concentrations. Thus higher concentrations produce more uniform fibres compared with lower concentrations. Various standard deviations support this finding. The beads are found to be more prominent at lower concentrations as compared with higher concentrations. The viscosity of the polymer solution is influenced by the concentration. Thus, the polymer in acid has lesser molecules at low concentrations, which decreases the cohesion and the solution is easily blown by electrostatic forces producing an unstable electrospinning jet that results in poor fibre deposition and more broken fibres.

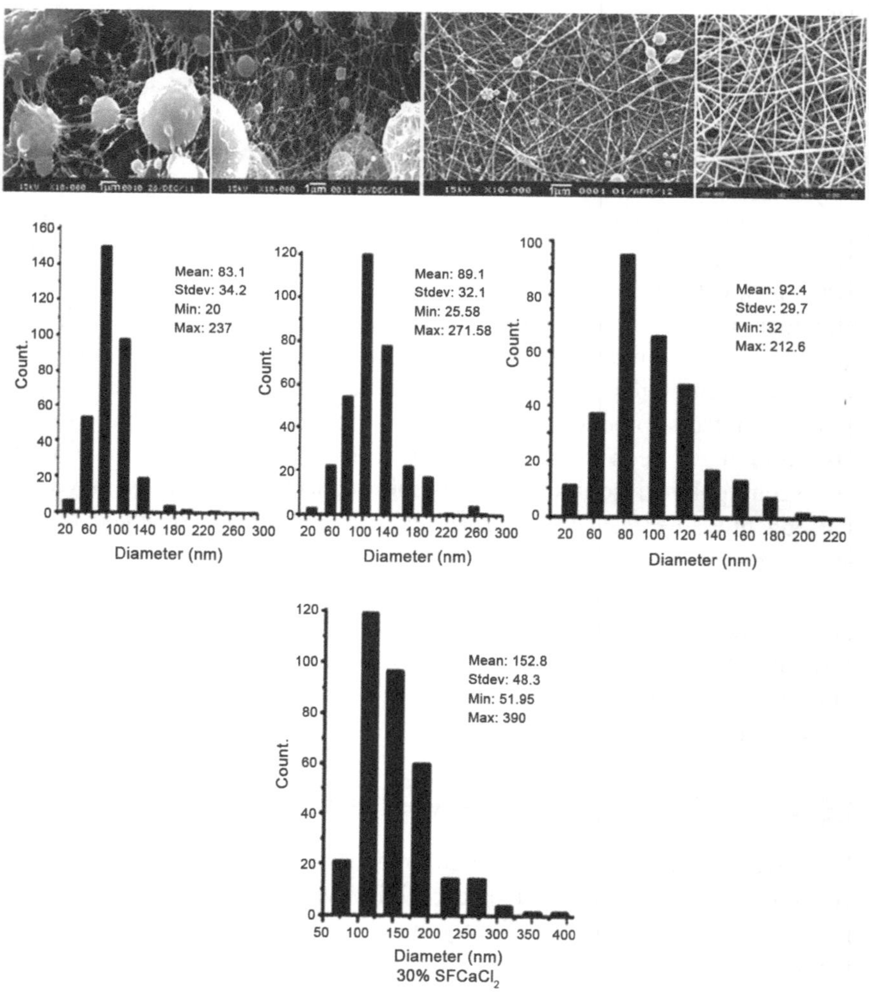

Figure 25.1 SEM photos and distribution of diameters of electrospun calcium chloride silk fibres (Bombyx mori) [13]

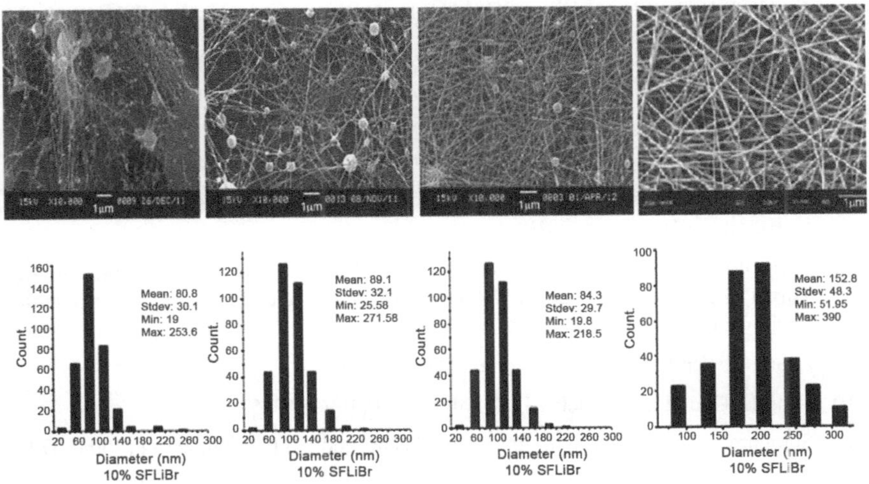

Figure 25.2 SEM photos and distribution of diameters of electrospun lithium bromide fibres (Bombyx mori) [13]

Besides, the polymer solution reaches the collecting plate before all the acid evaporates off, which promotes bead formation. Other factors which may influence formation of beads include collecting distances, flow rates and electric field among others. But, with such fixed factors, such findings reveal that higher concentrations enable better fibre lay and decreases occurrence of beads in the nano web. Generally, higher concentration exhibit closer fibre packing, fewer beads and more uniform webs than lower concentrations.

The fibre diameter has been measured and found to widely range between 19 nm and 394 mm. The mean diameter is found to increase with concentration, considering errors in experiment, testing and analysis. With the increase in concentration, the minimum fibre diameter also increases. At higher concentration, the time for the electrospinning jet thinning reduces. Hence, the diameter remains higher than observed in lower concentrations. The polymer molecules are small in number in the spinning solution and thus less interaction of molecules results and aggregate formation occurs during lower concentrations, which leads to lower fibre formation and increases bead formation.

References

1. Zarkoob S. Structure and morphology of regenerated silk nano fibers produced by electrospinning, PhD Thesis, University of Akron, 1998.
2. Zarkoob S, Renekar DH, Ertley D, Eby RK and Hudson SD. US Patent 6, 110,590, 2000.

3. Jin HJ, Fridrikh SV, Rutledge GC & Kaplan D. Biomacromolecules, 3(2002) 1233.
4. Zhang Q, Yan S & Li M. Materials, 2 (2009) 2276.
5. Ayutsede J. Regeneration of Bombyx mori silk nanofibres and nanocomposite fibrils by the electrospinning process, PhD Thesis, Drexel University, 2003.
6. Kim U, Park J, Kim JH, Wada M and Kaplan DL. Biomaterials, 26(15) (2005) 2775.
7. Meinel AJ. Cell instructive silk fibroin scaffolds for tissue engineering, PhD Thesis, University of Wurzburg, 2010.
8. Cappello J and McGrath KP. In Silk Polymers, Materials Science and Biotechnology, Vol.544, edited by D Kaplan, WW Adams, B Farmer and C Viney (American Chemical Society), 1994.
9. Zhang Q, Yan S and Li M. Materials, 2 (2009) 2276.
10. Fong H, Chun I and Renekar DH. Polymer, 40(1999) 4585.
11. Sukigara S, Gandhi M, Ayutsede J, Micklus M, and Ko F. Polymer, 44 (2003) 5721.
12. Deltzel JM, Kleinmeyer J, Harris D and Tan BNC. Polymer, 42 (2001) 261.
13. Kamalha E, Zheng YS, Zeng YC and Mwasiagi JI. Effect of solvent concentration on morphology of electrospun Bombyx mori silk, Indian Journal of Fibres and Textile Research, Vol. 39 (2004) 201.

26

Technical aspects of silk sericin

Summary : It has been found that the quality of sericin quality varies with the source and the method of extraction. The sericin derived from woven silk fabric is found to be the best in quality and quantity, whereas the sericin derived from silk waste is the poorest. Owing to greater and faster extraction, IR heated machine is found to be far more efficient when compared with HTHP machine. The drying technique influences the morphology of the surface. Globular aggregated particles are obtained by spray drying, whereas planar sheet-like particles are obtained by freeze drying. Based on the source or extraction technique, molecular weight of sericin can range between 20 kDa and 200 kDa. Sericin having low molecular weight dissolves in water easily, whereas that having greater molecular weight is soluble only in hot water. Also it suffers damage to a certain extent during extraction as revealed by spectroscopic and other methods of analysis. It is necessary to determine the influence of this degradation on the performance properties. Such findings can enable evaluation of the sericin protein quality derived from different sources or methods of extraction and processing.

26.1 Introduction

Chemically, silk comprises of fibroin and sericin, along with very small amounts of waxy substances, mineral salts, and coloring matter [1]. Both sericin and fibroin are proteins. While sericin contributes to nearly 75% of the weight silk, fibroin contributes to the remaining 25% [2]. Basically sericin is amorphous, more water soluble and plays the role of gum binder to maintain the structural integrity of the cocoon, and is rich in serine (about 32%), aspartic acid (16.8%) and glycine (8.8%), and also has hydroxyl groups in a high concentration [3, 4]. For long time, it was treated as a waste byproduct and normally discarded. However, recently persistent attempts have been made to recover and reuse it as a natural biopolymer in different areas of applications. It is interesting to note that sericin exhibits several important useful properties, like antioxidant, anti-tyrosinase, UV absorbing and moisture absorbing properties, as pointed out by recent researches [5–7]. It finds applications in the following areas:

(a) Finishing agent for natural or manmade textiles
(b) In the cosmetics industry as skin care products
(c) In the food industry as a valuable natural ingredient
(d) Along with fibroin it is used for biomedical polymeric applications [8,9]

DOI: 10.1201/9781003735427-26

In the case of silk degumming, a number of processes like extraction with water at high temperature and high pressure, boiling off in soap, alkalis, acidic solutions, aliphatic amines, enzymes and bio-surfactants have been adopted [10–15]. Efforts have been taken to recover sericin from soap alkali bath by the use of micro, ultra and nano filtration methods [16]. Methods such as spray, freeze or tray drying have been used for drying [17–19]. In the case of lab-scale applications, other methods of purification like precipitation by ethanol have been used [20]. The method of sericin extraction and recovery influence the properties [21]. During degumming process, sericin is fully/partially hydrolyzed, and solubilized in degumming medium. Hence, it is necessary to develop extraction processes for sericin that are energy efficient, produce no chemical discharge and cause minimum damage to sericin. With regard to laboratory applications, various types of energies are available during recent times. Infra red dyeing and drying machine works on the principle of radiation heating where heat is transferred in the form of electromagnetic waves directly to the center of material without heating the surrounding air [21]. The IR heating has gained greater acceptance since it has higher thermal efficiency and faster heating rate than convection heating [22]. Since the extraction conditions can affect the final properties of sericin, it is necessary to have a comprehensive analysis of the same. As sericin is easily affected by heat, the extraction conditions need proper monitoring. The quality and efficiency of sericin extraction are influenced by factors like temperature and duration time of treatment. The economical aspect of the method of sericin extraction is determined by the yield. Sericin has been extracted from four sources of mulberry silk; and the influences of different factors on sericin extraction have been investigated by means of infrared dyeing and HTHP machines. The sericin extracted has been compared with standard one and also has been characterized.

26.2 Technical details

In order to extract sericin, the silk obtained from different sources have been used. These include silk cocoon, flat silk, reeling silk waste, and woven silk fabric, and represent different stages of silk processing and hence have varying levels of purity and quantity of sericin [23]. The purest form of silk exists as the silk cocoon, but is not common and also expensive. As silk fabrics undergo degumming process for removal of sericin, these are the most obvious and abundant source for collection of sericin. Sericin has been extracted from *Bombyx mori* silk [37]. Analytical grade chemicals used are sodium carbonate, sodium hydroxide, copper sulphate, sodium potassium tartarate, bovine serum albumin and Folin ciocaleau's phenol reagent, and these have been used for protein estimation. Deionized water has been used for all extractions.

The study has been carried out in the following stages:

(a) Extraction of sericin [23]
(b) Determination of quality and quantity of sericin [24]
(c) Preparation and characterization of sericin powder [25]
(d) X-ray diffraction
(e) Spectroscopic analysis
(f) Fluorescence spectroscopy
(g) Circular dichroism spectroscopy
(h) Thermogravimetric and SEM study
(g) SDS-PAGE analysis of sericin [34]

26.3 Analysis of extracted sericin

Cocoons, flats, silk waste and woven fabric are the sources of extraction of sericin. Spectroscopy and determination of protein content have been used in the qualitative evaluation of the liquor obtained from these extractions. The results have been benchmarked against standard sericin obtained from sigma.

UV spectroscopy

Figure 26.1 depicts the UV spectra of standard sericin and sericin extracted from different sources. In the case of standard sericin and test samples, the characteristic peak due to the absorption of amino acids is observed at 275.40 nm. The presence of peptide bonds and aromatic acids enable the proteins to strongly absorb the sericin based on ultraviolet region. The aromatic amino acids like tryptophan, tyrosine and phenylalanine are known to absorb UV radiation in the range 260–290 nm due to $\pi \geq \pi^*$ transition, and these absorptions can be used to assess the quality of sericin protein.

Different samples exhibit different shape of the absorption spectrum, despite all samples of sericin showing the characteristic protein peak at 275.4 nm [37]. There is a broadening of peak in the sericin obtained from the cocoons, silk flats and silk waste in comparison with the standard sericin. This implies certain changes in the properties of protein. The sericin extracted from fabric exhibits a spectrum nearest to that of the standard. This trend is endorsed by the values of yield (%) and A-ratio (A_{290}/A_{260}) for various samples. In the case of silk fabric, highest yield of ~ 28% has been noticed. In the case of silk flats and cocoons, slightly less yield is achieved. With silk waste, lowest yield of ~22% is obtained. The A-ratio of sericin extracted from various sources differs for same extraction conditions. This shows that the extraction source greatly influences the sericin quality. The order of source based on A-ratio is as follows:

Standard sericin > fabric > cocoons > flats > waste

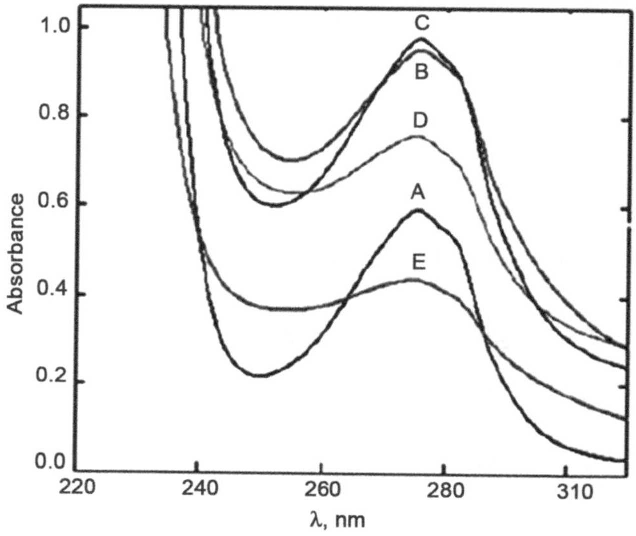

Figure 26.1 UV spectra of sericin obtained from various sources [37]
A – Standard sericin; B – Cocoons; C – Fabric; D – Flats; and E – Waste

The highest A-ratio of 1.73 is exhibited by the standard sericin sample and falls near to the ideal value of 1.8. This is followed by sericin extracted from silk fabric with a value of 1.35. The sericin quality improves with higher value of the A-ratio. In the case of sericin extracted from cocoons, the value did not match the ideal one. The least value of A-ratio of 1.09 is found in sericin extracted from silk waste, which shows that during silk reeling process, certain changes or damage occur in sericin [37]. It gives a broad peak in UV and a corresponding low value of A-ratio. Hence, among all the sources, silk fabric is found to give the maximum yield and highest quality of protein.

Protein content

Protein content of the sericin extracted from different sources has been obtained. It is not possible to correlate the results since standard sericin is available in powder form, while other samples are available in the form of liquor. In the case of silk fabric, the protein content in extracted liquor is highest at 31.62 mg/mL, while the values of protein content are 29.12 mg/mL and 28.2 mg/mL, respectively, in liquors extracted from fabric and cocoons. In the case of extracted liquor from silk waste, the protein content is lowest at 23.2 mg/mL [37]. Hence, considering the quality as well as quantity of sericin extracted from different sources, the best source is found to be from fabric. Since silk fabric is the most easily and abundantly available source as compared to the others, it is decided to conduct all further studies with this source.

26.4 Sericin extraction with various energy sources

Soap and alkali are used in conventional degumming of silk. The sericin protein degrades due to alkali. But, as the method is economical and sericin recovery is not a consideration for many processors, the industry continues to adopt it. As the method results in addition of large quantities of soap and alkali into the effluent stream, it is highly polluting [37]. More critically, it is extremely difficult to recover sericin from these effluent streams. In order to maximize the recovery of sericin and minimize the damage to protein, extraction by water is the best option. HTHP machines have been used to separate sericin from the fabric [26]. But, certain recent investigations on sericin extraction from silk waste indicate that HTHP extraction is not effective and causes damage to the protein [27]. Infrared heating has been used for the extraction of sericin from silk fabric and results are compared with HTHP method.

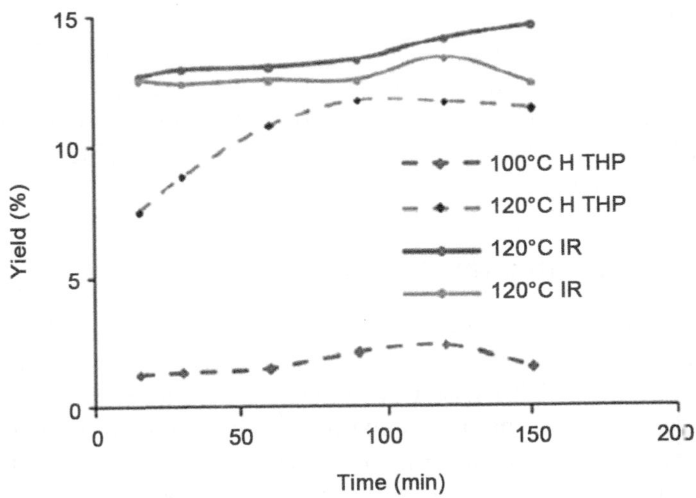

Figure 26.2 Influence of temperature and time duration [37]

The sericin yield due to extraction from fabric for various time and temperature of extraction using IR and HTHP machine have been determined. The sericin yield at 15 min duration of extraction in IR method at 100°C is found to be ~12.6%. With sericin yield increasing up to ~14.6% at 150 min increase in time, the yield is found to increase slightly. There is no improvement in the sericin yield with further increase in temperature at 120°C. But, with HTHP extraction, the trend noticed quite different. When compared with IR method, the yield achieved in all cases is much lower. In the case of HTHP method, the influence of temperature is much more prominent than that of IR heating bath. Despite extraction duration of 120 min, the yield at 100°C is very low (~3%). As the time and temperature of extraction increases, the yield also increases. Maximum yield of ~12% is obtained at 120°C for 90 min of extraction. These observations clearly show that IR heating is a much more efficient method of sericin extraction as compared to HTHP heating. There is higher yield, requiring less time and temperature. Such a result falls in line with earlier findings [23]. Possibly, the heating principle used in both methods could create the difference in yield from IR to HTHP method. As the bulk of sericin is removed within 15 min, it is indicative of the fact that IR is much more efficient than HTHP in this regard. Since sericin, like all proteins, is prone to degradation at high temperatures, 15 min at 100°C is taken as the suitable procedure for the extraction of sericin from silk fabric. The UV spectra of standard sericin and of that extracted by HTHP and IR machine show the characteristic peak at 275.4 nm. However, slight peak broadening is observed from samples B and C, thus suggesting some changes in the sample. Also, higher A-ratio (1.350) which is indicative of protein quality is observed for the sericin sample extracted in IR machine as compared to that extracted in HTHP. The findings indicate that the source of energy/the method of heating critically influences the quality of protein obtained. The test results indicate that extraction of sericin in IR machine 100°C for 15 min is found to be the best method of extraction, as lower temperature is desirable for avoiding denaturation of proteins.

26.5 Sericin characterization

Sericin superior in quality and quantity is obtained when extracted from fabric at 100°C for 15 min using infrared heating. Hence, it is changed from liquor to powder state, and the characterization data relating to its composition, spectroscopic, thermal properties and molecular weight are generated. The findings are benchmarked against the standard sericin sample and discussed herein [37]. The sericin powder extracted from fabric and the standard sericin sample are characterized in terms of moisture, ash, nitrogen and protein content. Moisture content of the standard sericin sample is found to be slightly higher than the sericin extracted from the fabric. This may be because a different method of extraction may have been used for the standard, resulting in higher residual matter. Gulrajani et al. have reported that ash content can vary from 0.8% to 5.2% in sericin extracted from cocoons using HTHP and alkali method, respectively [18]. Wu et al. report an ash content of 4.2% in sericin extracted from cocoons using HTHP method. The nitrogen content of the sericin powder is also estimated and the protein content is calculated by multiplying the value by 6.25. There greater nitrogen content (14.13%) and thus greater protein content (88.31%) in the sericin sample obtained from the fabric, than that of the standard sericin sample, and is also similar to the values (14.65% nitrogen) reported earlier [28].

26.6 X-ray diffraction analysis

The crystalline of sericin samples have been investigated by means of X-ray diffraction. In the case of silk protein three kinds of conformations have been proposed [28, 29, 37]. Silk I is referred to as the glandular state before crystallization. The spun silk state that comprises of the β-sheet secondary structure is referred to as silk II, and silk III (an air/water assembled interfacial silk) is a helical structure. The main diffraction peaks of silk I are present at around $2\theta = 12.2°$ and $28.2°$, while silk II are present at about $2\theta = 18.9°$ and $20.7°$. The X-ray diffraction curves of sigma sericin and sericin extracted from fabric exhibit a strong diffraction peak around $2\theta = 20.5°$. The powders are amorphous since it has been noticed that the peaks are broad. In all the sericin samples including the reference one, the X-ray diffractograms are observed to be the same. Earlier investigations have reported similar findings [28].

26.7 FTIR analysis

The proteins exhibit characteristic vibration bands between 1630 cm^{-1} and 1650 cm^{-1} for amide I (C-O stretching) and 1540–1520 cm^{-1} for amide III (C-N and N-H functionalities), as seen from FTIR spectra. Also, in the case of amides I, II and III, such band positions conform the protein materials, like 1650 cm^{-1} (random coil) and 1630 cm^{-1} (β-sheet); 1540 cm^{-1} (random coil) and 1520 cm^{-1} and 1230 cm^{-1} (random coil), respectively [37] Sericin extracted from fabric shows absorption between 1600 cm^{-1} and 1700 cm^{-1}, confirming amide I absorption which arises predominantly from the C=O stretching vibration and is most useful for determining proteins secondary structure. Due to the random coil structure the peak of sericin arises at 1540 cm^{-1} for amide II. With both sericin samples, the signature peak has been noticed at 1400 cm^{-1}. Also, the characteristic peak for the amide III occurs between 1240 cm^{-1} and 1250 cm^{-1} mainly due to the C-H stretching vibration coupled to the N-H plane bending vibration, and is observed to shift in the range corresponding to a change from random coil conformation to β-sheet structure. In the case of IR spectra obtained from standard and test samples extracted from various sources and prepared by various techniques, no greater difference has been noticed. The results are same as those reported by other research workers [30, 31].

26.8 Fluorescence spectra

In order to determine whether extraction parameters influence the conformational state of sericin in anyway, fluorescence spectra of the standard and test samples have been used Fig. 26.3(a). At 355 nm, the standard sericin (A) exhibits a distinct peak, while sericin derived from fabric (B) exhibits a minor peak at around the same wavelength and a distinct peak at 380 nm [37]. The fluorescence of a folded protein is a mixture of the fluorescence from individual aromatic residues. The excitation of tryptophan is responsible for most of the intrinsic fluorescence emissions of a folded protein, with some emissions arising from tyrosine and phenylalanine. Typically, tryptophan has the excitation wavelength of 295 nm and an emission peak that is solvatochromic, ranging from 300 nm to 350 nm depending on the polarity of the local environment [32]. In Fig. 26.3(a), the extracted sericin sample shows a fluorescence pattern which is different from that of the standard sample. This indicates that some changes do take place in the sericin, during extraction processes that affect the conformational state of sericin protein. It is necessary to establish the influence of such changes on the biological and performance properties like as cell regeneration, antioxidant property and moisture absorption needs for understanding the relevance of the structure change.

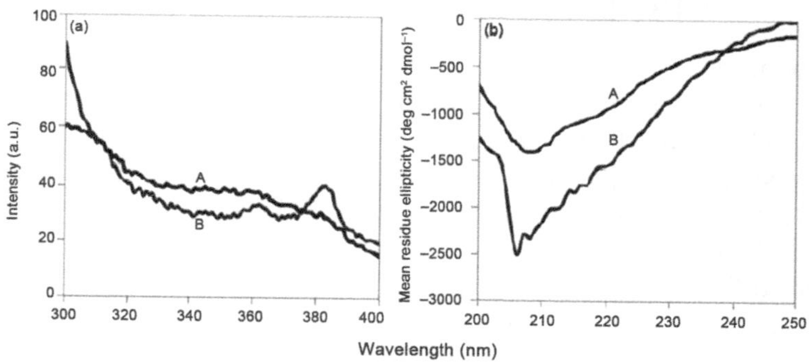

Figure 26.3 Sericin samples showing (a) fluorescence spectra and (b) CD spectra [37]
A – Standard sericin, B – Sericin from silk fabric

26.9 Circular dichroism (CD) spectra

The secondary structure of proteins in the far UV spectral region (190–250 nm) can be determined by CD spectroscopy, which measures the differences in the absorption of the left-handed polarized light against right-handed polarized light [37]. At these wavelengths, a chromophore is a peptide bond, and a signal arises, when it is located in a regular-folded environment, α-helix, β-sheet and random coil structures that provides characteristic shape and magnitude of a CD spectrum. At 208 nm, the negative peak characterizes an α-helix protein, and at 214 nm the negative peak characterizes β-sheet of protein [27]. In the investigation, CD curves (Fig. 26.3b) of the IR-extracted sericin sample exhibits a negative band at 206–208 nm indicating α-helix conformation. It has been found that at 198 nm sericin recovered from HTHP degumming liquor exhibits a strong negative band and at 218 nm a weak band which is indicative of random coil and β-sheet configuration, respectively. On the other hand, sericin recovered from alkaline degumming shows a negative peak at 201 and 216 nm, revealing the presence of α-helix structure [18]. Similar results have also been reported for the secondary structure of sericin prepared by ethanol precipitation. Low molecular weight sericin recovered from *A. mylitta* also shows similar peaks in its CD spectra [33]. Based on the extraction method, it can be concluded from these observations that sericin can show both α-helix and β-sheet structure.

26.10 Molecular weight

SDS-PAGE technique has been used to determine the molecular weight distribution of the various sericin samples, and the results are depicted in Figure 26.4. There is a diffused band in the molecular weight range 6.5–205 kDa with the sericin extracted from fabric, whereas it is 3.5–43 kDa with standard sericin. In the case of extracted silk, the molecular weight is much higher than that of standard sericin [37]. Also, the solubility of the two samples differs when dissolved in water [27].

Standard sericin having lower molecular mass (< 50 kDa) is easily soluble in cold water, whereas extracted sericin having a higher molecular mass between 50 kDa and 200 kDa only be dissolved in hot water (90°C). This large difference in molecular weight can have a significant effect on the properties and applications of sericin. It has been reported that sericin consists of a group of protein molecules of molecular weight ranging from 20 kDa to 400 kDa. One research reported that the molecular mass of sericin is 309, 177, 145, 134 and 80 kDa, while another reported at least 15 different polypeptides with molecular weight ranging between 20 kDa and 200 kDa in the anterior portion of middle portion of middle silk gland [34, 35]. Research has revealed that sericin appears in a continuous distribution in the range 97–14 kDa, while some bands are observed above and below 97 kDa, respectively.

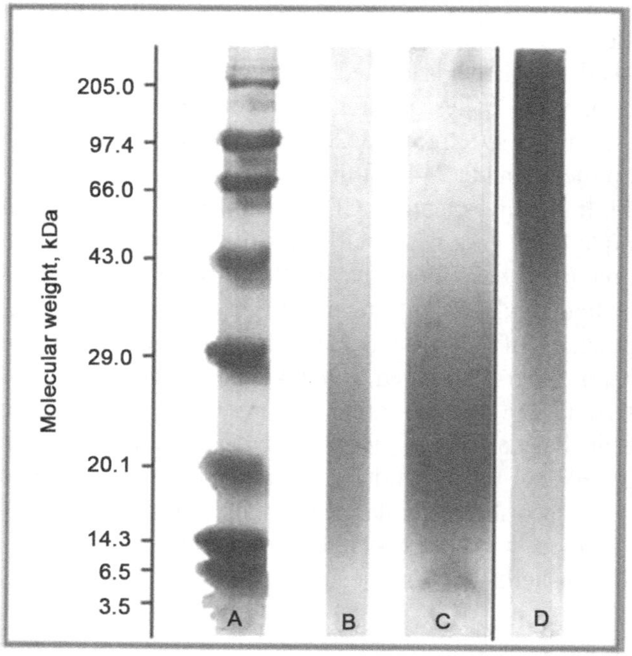

Figure 26.4 SDS page analysis of sericin samples [37]
(A) Marker, (B) Standard 20 mg/mL, (C) Standard 60 mg/mL,
(D) Sericin from fabric 10 mg/mL

26.11 TGA

It has been used to analyze the thermal behavior of sericin. There has been initial weight loss due to evaporation of water as shown by the weight loss trace [27]. Subsequently, a sudden reduction in mass has been seen in the wide temperature range above 220°C in both the samples. Similar trend has been reported for sericin [36, 37]. It is clear from the weight-loss pattern in the thermograms that sericin extracted from the fabric exhibits higher weight loss as compared to the standard sericin, thus indicating that the sericin prepared from fabric is relatively unstable to temperature. In the case of standard sericin powder, greater residue is obtained in standard sericin powder (9%) than that of the sericin powder recovered from the fabric (5%). The findings agree well with % ash content as reported above.

26.12 Investigations on morphology

SEM has been used to investigate the morphology of sericin powders. The shape of the particle differed much in the two cases as observed from the micrographs (Fig. 26.5). There are spherical globules in the test sample, while the standard sample comprises of long planar size particles. Also, the spherical globules have been found to have a very wide size distribution and appear greatly aggregated. These particles are generally not easily soluble [27, 37]. To study if the particle shape is indeed characteristic of a drying technique, another test was conducted where the extracted liquor was freeze dried instead of spray drying. The SEM of the powder obtained by freeze drying is given in Fig. 26.5(c). Planar-shaped particles are obtained indicating that particle shape is a function of the method of drying. The properties of the globular particles are found similar to all other properties of the new powder.

(a) (b) (c)

Figure 26.5 SEM photos of sericin samples [37]
A – Standard sample, B – Spray dried sample, C – Freeze dried sample

References

1. Teli MD and Rane VM. Fibres and Textiles Eastern Europe, 19(2011) 10.
2. Freddi G, Mossotti R and Innocenti R. Journal of Biotechnology, 106(2003) 101.
3. Chopra S and Gulrajani ML. Indian Journal of Fibres and Textile Research, 19(1994) 76.
4. Cho KY, Moon JY, Lee YW, Lee KG, Yeo JH, Kweon HY, Kim KH and Cho CS. International Journal of Biology and Macromolecules, 32(2003) 36.
5. Yun H, Oh H, Kima Mk, Kwak HW, Lee JY, Um IC, Vootlad SK and Lee KH. International Journal of Biology and Macromolecules, 52(2013) 59.
6. Sarovart S, Sudatis B, Meesilpa P, Grady BP and Magaraphan R. Reviews in advanced material science, 5(2003) 193.
7. Aranwit P, Damrongsakkul S, Kanokpamont S and Srichana T. Biotechnology and Biochemistry Research, 55(2010) 91.

8. Sasaki M, Yarmada H and Kato N. Numerical research, 20(2000) 1505.
9. Ki CS and Park reYH, Macromolecule Research, 17(2009) 935.
10. Gulrajani ML, Gupta SV, Gupta A and Suri M. Indian Journal of Fibre and Textile Research, 21(1996) 270.
11. Bianchi AS and Colomma GM. Melliand Textilbert, 73(1992) 68.
12. Rangan R, Wang L, Kanwar JR and Wang X. Journal of Applied Polymer Science, 119(2011) 1339.
13. Gulrajani ML, Sethi S and Gupta S. Journal of Society of Dyers and Colourists, 109(1993) 256.
14. Gulrajani ML and Sinha S. Journal of Society of Dyers and Colorists, 108(1992) 79.
15. Sarma M, Gogoi S, Devi D and Goswami B. Journal of Scientific and Industrial Research, 71(2012) 270.
16. Capar G, Aygun SS and Gecit MR. Journal of Membrane Science, 325(2008) 920.
17. Gen CG, Bayratkar O and Basal G. Tekstil ve Konfeksiyon, 19(4)(2009) 273.
18. Gulrajani ML, Purwar R, Prasad RK and Joshi M. Journal of Applied Polymer Science, 113(2009) 2796.
19. Vaithanomsat P and Kitpreechavanich V. Separation and purification technology, 59(2008) 129.
20. Kurioka A, Kurioka F and Yamazaki M. Bioscience, biotechnology and biochemistry, 68(4)(2004) 774.
21. Mandal BB, Ghosh B and Kundu SC. International Journal of Biology and Macromolecule, 49(2011) 125.
22. Krishnamurthy K, Khurana HK, Jun S, Irudayaraj J and Demirci A. Comprehensive Review of Food Science and Food Safety, 7(2008) 1.
23. Gupta D, Agrawal A, Chaudhary H, Gupta C and Gulrajani M. Journal of Cleaner Production, 52(2013) 488.
24. Boucau J. Enzymatic and structural characterization linked to mycobacterium tuberculosis pathogenically of proteins, PhD Thesis, The University of Toledo, 2008, 39.
25. Official method of analysis of AOAC international, 17th edition (Association of official analytical chemists, Gaithersburg, MD) 2000.
26. Li G, Liu H, Li T and Wang. Journal of Material Science C, 32(2012) 627.
27. Aramwit P, Siritientong T and Srichana T. Waspe Manage Research, 30(3) (2012) 217.
28. Wu JH, Wang Z and Xu SY, Food Chemistry, 103(2007) 1255.
29. Lamoolphak W, Eknamkul WD and Shotpruk A. Bioresource Technology, 99(2008) 7678.
30. Teramoto H and Miyazawa M. Biomacromolecules, 2049.
31. Zhang XM and Wyeth P. Science China, 53(2010) 626.

32. Lackowicz JR. Protein fluorescence, in Topics in Fluorescence spectroscopy, Vol.6 (Kluwer Academic publishers, New York), 2002.

33. Dash R, Ghosh SK, Kaplan DL and Kundu SC. Comparative Biochemistry and Physiology, Part B, 147(2007) 129.

34. Gamo T, Inokuchi T and Laufer H. Insect biochemistry, 7(1977) 285.

35. Sperague K. Biochemistry, 14(1975) 925.

36. Tsukada M. Journal of Applied Polymer Science, 22(1978) 543.

37. Deepti G, Anjali A and Abhilasha R. Extraction and characterization of silk sericin, Indian Journal of Fibres and Textile Research, (2014) 364.

27

Application of genetic algorithm for design of silk scaffolds

Summary : Neural network has been used to develop model for the prediction of the final mechanical properties of scaffold after extracting and visualizing the main features of the data set. The mechanical behavior has been analyzed adopting the desirability approach as the basis. The main characteristics have been optimized using genetic algorithm technique. Such techniques have enabled to establish a suitable model which can predict and optimize the global mechanical properties of wire-rope yarn. As an instance, the anterior cruciate ligament (ACL) having the best structure for young people is 2-5-4-5 with regard to number of filaments and 57-36-46-31 with regard to number of twist in each layer of wire rope scaffold, respectively.

27.1 Introduction

In the area of tissue engineering, scaffolds help to give an initial support and framework for attaching, profilerating and differentiating various cell lines as an extracellular matrix [1]. In orderly effectively design tissues such as tendon and ligament in the area of tissue engineering, it is necessary that a typical scaffold should be able to provide biological signals coupled with suitable mechanical behavior [2]. In the construction of scaffolds, suitable mechanical support becomes an essential consideration. Hence, different techniques have been evolved to further improve the mechanical performance of scaffolds [3]. Scaffolds made of fibre have been widely used to simulate tendon and ligament tissues owing to proper mechanical properties [4]. Such tissues possess unique mechanical properties such as viscoelastic and non-linear behavior which are the same as fibre base structures [5, 6]. Also, textile structures like woven, knitted, braided or twisted types made of fibres are almost similar to tendon and ligament tissues, and hence can be the best option to design and fabricate their scaffold. Silk is a material that combines high strength (up to 4.8 GPa), remarkable toughness, elasticity (up to 35%), and environmental stability. The potential of native silk fibroin fibres have at first been explored as three-dimensional scaffolds for tissue engineering of the anterior cruciate ligament (ACL). Mechanical properties of human ACL are comparable to textile structure as reported in a study as a twisted structure [7]. Because of the importance of simulation mechanical properties, recently

DOI: 10.1201/9781003735427-27

researchers have attempted to develop the relation between structural and mechanical of a scaffold by various methods like mathematical or statistical method precisely. Artificial neural network technique has been used in the investigation to evolve a new method for the determination of mechanical properties of silk wire-rope scaffold used in tendon and ligament tissue engineering. For the first time, a genetic algorithm model is developed for the automation and optimization of scaffold designing. The ANN is one of the most powerful information processing systems that imitate the function of the human brain and biological neural networks. This technique is useful when there are a large number of effective parameters in the special process without requiring a prior knowledge of the relationships of process factors [8]. ANNs comprise of simple processing elements, known as artificial neurons [9]. One of the models that is generally used among various types of ANN is the multi-layer perception (MLP) neural networks with back propagation (BP) training procedure for modeling a problem [10, 11]. Neural network models are most effective for prediction and simulation of structures so as to understand the influence of some production parameters in different applications such as tissue engineering, from the scientific perspective [12]. Also, genetic algorithms (GAs) have been widely and successfully applied to various optimization processes [13]. GAs are well appropriate for the concurrent manipulation of models with changing solutions and structures, since they can search non-linear resolution spaces without requiring gradient information and prior knowledge about model features [14]. Based on the extent of environmental adaptation of chromosomes, the fineness value is taken as the basis of ranking chromosomes, so as to understand which chromosome would survive in the next generation. Such techniques are powerful tools for finding the target optimal formula in optimization problems. By the self-adaptation function and threshold, the system itself has the ability to evolve through optimum solution for the problem [15]. GA has been used to optimize effective parameters that result in optimum mechanical properties due to its intuitiveness, ease of implementation and its ability to effectively solve highly nonlinear and mixed integer optimization problems to optimize effective parameters in the investigation for fabricating a silk wire-rope scaffold [16]. In the proposed GA model, the effective parameters of silk wire-rope scaffold in tendon and ligament tissue engineering are dynamically optimized by implementing the GA evolutionary process and then performing the prediction task using these optimal values. The optimal values of parameters are searched by Gas with a randomly generated initial populations consisting of chromosomes. The values of the two parameters, namely the number of filament and the number of twist in each layer of wire-rope yarn, are directly coded in the chromosomes with real-valued data. The single best chromosome in each generation is the survivor of the succeeding generation. In the field of tissue engineering, the main research problem has been the prediction of mechanical properties of scaffold. Hence, the GA model being more accurate than the traditional multivariate statistical models and neural network method, has been applied to the problem in preparing scaffolds by tissue engineering to verify its accuracy and generalization ability.

27.2 Technical details

Silk hierarchical structure scaffolds have been designed that resemble arrangement of collagen fibres in tendon and ligament tissue. It is necessary to make cautious prior selection for values of parameters in wire-rope scaffold, so as to design an effective model [17, 22]. Such parameters comprise number of filaments in each layer (parameters for first to fourth layers, respectively), that determine elongation at break in a wire rope scaffold. There is variation in number of filaments and twist in each layer. Scaffolds have been identified by the type of structure, by the number of filament, and by the number of twists in various layers. The wire rope scaffold structure is schematically depicted in Figure 27.1.

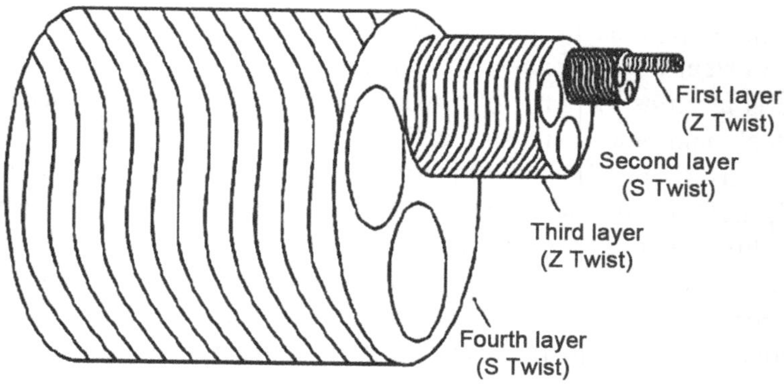

Figure 27.1 Wire rope scaffold for tendon and ligament tissue engineering [22]

Based on Taguchi orthogonal matrix, the data for ANN and GA model have been collected from a certain number of silk wire-rope scaffold samples. A single yarn of specific denier, ultimate tensile strength and elongation at break has been determined after degumming process. In the case of silk medical application, sericin layer has to be removed. Thus, degumming bath of sodium carbonate solution at 95°C for half hour and repetition of the process for one more hour has been carried out in a fresh degumming solution [18]. Just after setting final twisting, a number of various scaffolds have been characterized by mechanical properties analyzing samples in each group.

The whole experimental work comprised of following steps:

(a) Sample preparation and experimental design
(b) Modeling and methodology [19]
(c) Neural network model for prediction of features of ACL
(d) Application of GA for optimized structure [20, 21]

27.3 Findings of the study

In order to predict the mechanical properties of wire-rope hierarchical scaffold in applications of tendon and ligament tissue engineering, a model has been designed by application of the ANN. The input values in the model have then been optimized with GA, and comprised of the number of filament and the number of twist to receive the best structure [22]. First of all, the initial population size is optimized. With a large population size, the genetic algorithm searches the solution space more thoroughly, thereby reducing the chance that the algorithm will return a local minimum that is not a global minimum. But, the algorithm runs more slowly when the population size is large. As best and mean fitness function values are closing each other, all solutions are becoming similar to each other leading to one solution. The best fitness value normally gets enhanced after many generations and becomes approximately constant. In generation 30, convergence happens between mean and best values and the mean value is close to the best value, so optimum population size can define 30. The next parameter is the number of generations. For the groups of young, middle and old age, it was considered to have 30 generation.

Such values can be chosen for the optimum number of generation due to consideration of the time execution and convergence between mean and best values. There is significant growth of the execution time in the case of large values of generation number such as population size. After many trials, the other GA parameters along with population size and generation number have been set. In the young, middle and old age groups of all people, the mean and the best values fitness function values in each generation are depicted in Figure 27.3. In all figures for young, middle and old age, mean and the best values are very close and parameters setting in genetic algorithm model according are acceptable (Figure 27.3).

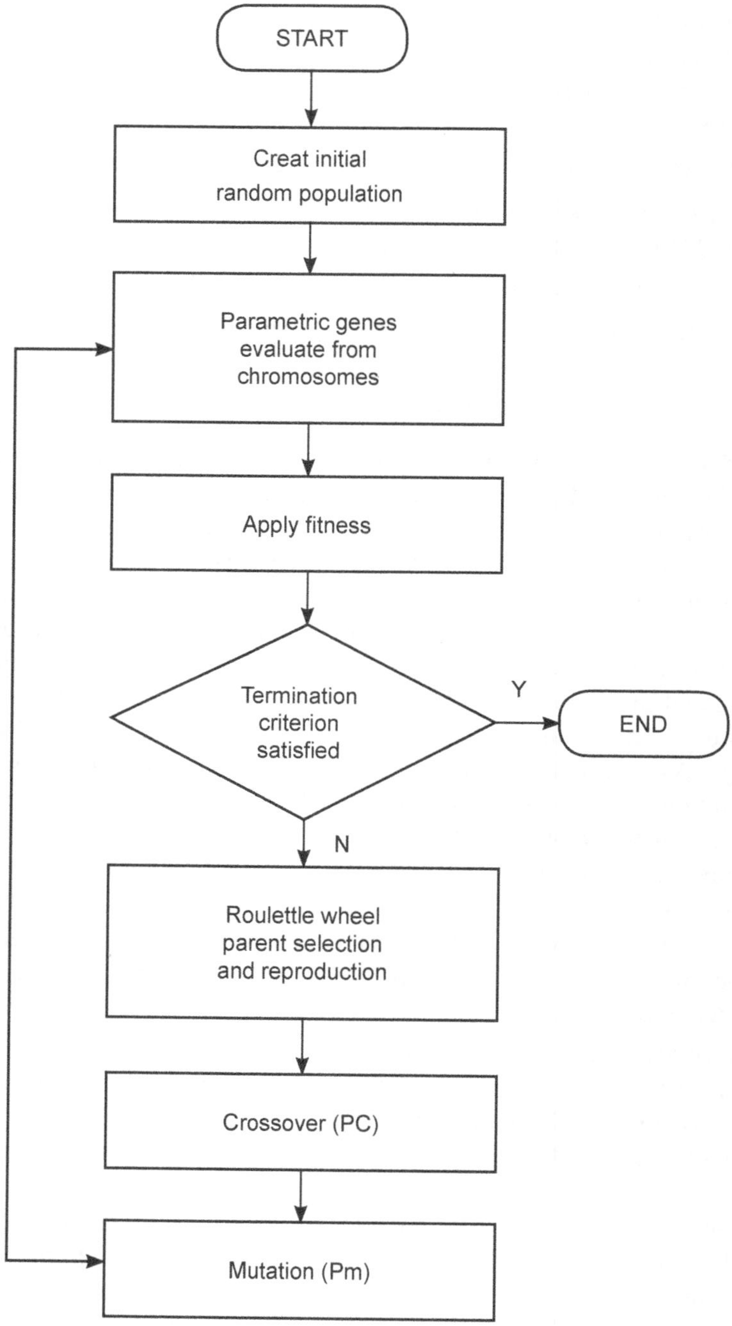

Figure 27.2 Structure of genetic algorithm [22]

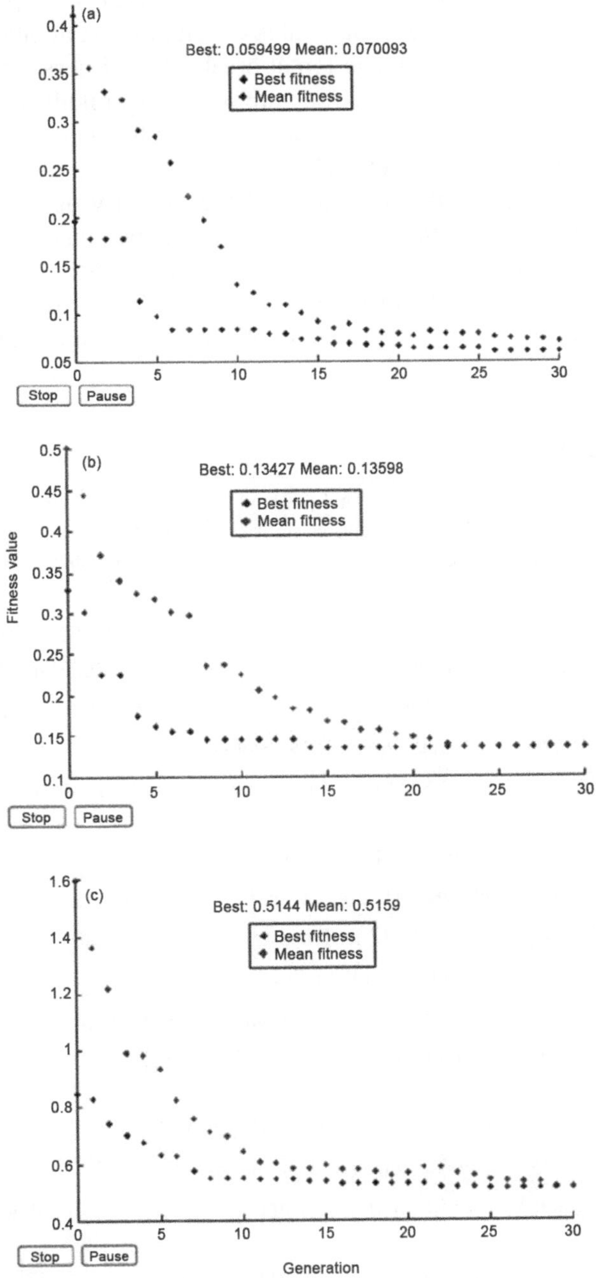

Figure 27.3 Fitness value vs. generation in optimization of genetic algorithm with regard to (a) young age group (b) middle age group and (c) old age group

After finding and running the best GA program, the optimization of input values for young, middle and old age including the number of filament and twist including UTS, elongation at break and stiffness were predicted. The finding reveals that the best structure for young, middle and old people is 2-5-4-5, 2-5-4-4 and 2-4-3-4 for the number of filament in each layers of wire-rope scaffold, respectively. The number of twist in each age group is also predicted along with the number of filament. By using these values, output parameters (mechanical behavior) are anticipated. Moreover, the prediction error for UTS and elongation at break changes from 10.7 to 13.65 as shown by the obtained data. Due to the interaction between UTS and elongation at break, the prediction error is much more in the case of stiffness.

References

1. Woo SLY, Almarza JA, Karouglu S, Liang R and Fischer MB. Functional tissue engineering of ligament and tendon injuries, Translational approaches in Tissue engineering and regenerative medicine (Artech house publisher), 2007, Chapter 9, 997.
2. Yang SH, Leong KF, Du ZH and Chua CHK. Tissue engineering, 7(2001) 679.
3. Hallister SJ, Maddox RD and Taboas JM. Biomaterials, 23(2002) 4095.
4. Karamuk E, Mayer J and Raebar G. Composite science technology, 64(2004) 885.
5. Daraski DM, Brink KS and Temenoff JS. Biomaterials, 28(2012) 187.
6. Cooper JA, Lu HH and Ko FK. Biomaterials, Mayer J and Raeber G, Composite science technology, 64(2004) 885.
7. Wang Y, Kim HJ, Vunjak-Novakovic G and Kaplan DL. Biomaterials, 27(2006) 6064.
8. Kumar AJP and Singh DKJ. Theoretical applied information technology (2006) 961.
9. Cirovic V and Aleksendric D. FME Transactions, 38(2010) 29.
10. Yaman N, Senol FM and Gurkan P. Engineered fibres and fabrics, 6(2011) 38.
11. Madle MJ and Marinkovic VJ. FME Transactions, 38(2010) 189.
12. Naghashzargar E, Semnani D, Karbasi S and Nekoee H. Textile Institute, 105(2014) 264.
13. Fogel DB. IEEE Transactions neural networks, 5(1994) 3.
14. McCall J and Petrovski A. Control and automation (1999) 65.
15. Semnani D and Vadood M. Engineering applications artificial intelligence, 23(2010) 217.
16. Vadood M, Semnani D and Morshed M. Applied polymer science,120(2011) 735.
17. Altman GH, Horan RL, Lu HH, Moreau J, Martin I, Richmond JC and Kaplan DL. Biomaterials, 23(2002) 4131.

18. Sahoo S, Toh SL and Guh JCH. Biomedical material research. Part B, Applied biomaterials, 95(2010) 19.

19. Cohl EPG and Vilensky ID. Textile Science – An explanation of fiber propert.es, 2nd edition (Guilford publications), 1984.

20. Woo SLY, Hallis JM, Adams DJ, Lynn RM and Takai Sh. American Journal of Sports Medicine, 19(1991) 217.

21. Lin Mt. International Journal of Clothing Science Technology, 20(2008) 258.

22. Elham N, Dariush S and Saeed K. Optimization of silk yarn hierarchical structure by genetic algorithm to design scaffolds, Indian Journal of Fibres and Textile Research, 40(2015) 81.

28

Special technique for the investigation of crystal and molecular structure of muga wild silk

Summary : X-ray diffraction data coupled with linked-atom-least-squares method have been used to investigate the crystal and molecule structure of muga wild silk fibre, belonging to Antheraea assamensis family. Four molecular chains are contained in the rectangular unit cell having parameters a = 9.44 A, b = 10.64 A and c = 6.96 A and the space group being P212121. Specific twist and shearing of the molecular chain is found along the fibre axis in muga wild silk fibre on comparing the crystal structure of silk II. In the case of muga silk fibre, it has been found that the sheet structures formed by hydrogen bonds assume the antipolar–antiparallel arrangement based on the obtained values of torsional angles, Eulerian angles and other parameters. This confirms the earlier findings. It is observed that the stabilization of the model comes from alternate arrangement of bonds C=O in the neighboring chains which results in a net small dipole moment.

28.1 Introduction

Wild silk is produced by anything other than mulberry silkworm. It is generally produced by caterpillars. There are a number of types of wild silks depending on the worm that produce them. Such wild silks are popular in countries like China, Europe and in few other southern parts of Asia for over many years. As such worms are not domestically cultivated, their yield is far lesser than that of the mulberry silk. As the cocoons are to be collected from the wild, they tend to get damaged by the time they are located due to the pupa leaving them, and hence results in shorter threads. Hence, the commercial utility of such silks is very limited and rare. Mulberry (*Bombyx mori*) and non-mulberry (Tassar, Eri and Muga) silks are the two major varieties. One of the wild silk varieties is the muga silk which is also known as *Antheraea Assamensis* (A.Assama). Few other wild silks are Mopani silk from South Africa, saturniidae silk from Thailand and assam silks (Muga, Eri and Pat) from India [1]. Sericulture generates employment for many people in India and one of the most economically important species for reared is the wild-type non-mulberry silkworm (*Antheraea mylitta*) [2, 3]. Muga wild silk has natural shimmering colour and is popular in the state of Assam. The crystal

DOI: 10.1201/9781003735427-28

and molecular structure of domestic and wild silk fibre varieties have been investigated by a number of researchers adopting different methods. Studies have been done on the crystal structure of *Bombyx mori* silk fibroin [4]. It is revealed that the silk is made up of regular arrangements of anti-parallel sheets. Amino acid sequence, basic building blocks, of silk fibres has been reported by number of researchers [5]. Investigation has been carried out relating to the refined molecular and crystal structure of silk I based on Ala-Gly peptide sequence [6]. Studies have also been reported on the crystal structure of pure Mysore silk (PMS) and the crystal and molecular structure raw bivoltine silk fibres [7, 8]. The comparison of parameters with pure Mysore silk fibre with other related research has been reported [4]. Another interesting area of research reported relates to crystal structure details of tassar silk fibres [9]. Since silk protein has proved very useful as a natural biopolymer in the area of tissue engineering and biomedical applications, its importance has increased. Even though it is not so familiar to the general public, it has attracted the interest of the researchers in this area who do work on the natural polymers and silks. The structure of some other silk species has been analyzed and continued attempt has been made in the topic discussed herein. This chapter highlights the effort taken to give emphasis on modifications in crystal and molecular structure of Muga wild silk (Indian) fibres using X-ray diffraction data and linked atom least squares (LALS) refinement modeling, so as to provide a better perspective of structure property relation in these fibres.

28.2 Technical details

The following stages have been involved in the investigation:

(a) Sample preparation

(b) Measurement by X-ray diffraction [10-12]

(c) Determination of structure based on the following considerations:

- Molecular model [13,14]
- Molecular and crystal structure for repeating unit

Fresh cocoons of muga wild silk obtained from germ plasma have been used. The cocoons have been cooked in boiling water under prescribed conditions, and then reeled in a mono cocoon reeling equipment. The fibres have been mounted in a frame in taut condition without mechanical stretching [17].

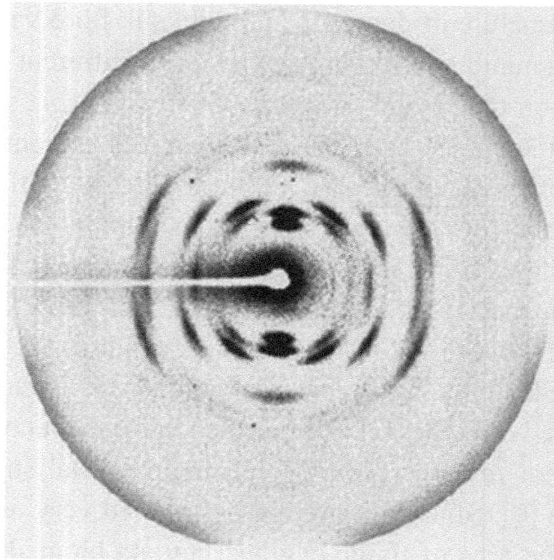

Figure 28.1 X-ray diffraction pattern of multivoltine wild silk fibre (corrected background) [17]

28.3 Findings of the investigation

The refined parameters and the final values have been determined. In this case, the azimuthal angles of the two molecules forming a sheet have also been separately refined. The comparison between the observed and calculated structure factors have been done [17]. The fractional atomic co-ordinates for (Ala-Gly) residues have been determined. Figure 28.2 depicts the crystal structure of muga silk in raw form. The glycine and alanine residues about N-C(=O) bonds have internal rotation angles of 15.04° and 177.16° respectively which are almost trans-conformation. The □ and Ψ for the glycine residue are ~153.87° and 149.86° respectively which are also between skew and trans-conformations. The □ and Ψ for the alanine residue are ~143.03° and 150.74° respectively, which are in agreement with the values (□ ≈ 146.67° and Ψ = 143.00°) given by Pauling and Corey for antiparallel-chain model [15,16]. Since the chain-a and chain-b are symmetric, their torsional and Eulerian angles are the same for both. In the case of muga silk fibre, the crystal structure along the fibre axis (i.e. c-axis) shows that the chains 2 and 2' are shared and twisted [Fig. 28.2(a)].

The stereo chemical energy which is represented by σ is found to be 2.37E+0.4 kcals. Here, the σ is given by the sum of second term of the Equation derived. It can be seen that the molecular modification is essentially same as β-pleated structure with antipolar–antiparallel arrangements formed by hydrogen bonds. In a unit cell, the position of the C=O of chain along c-axis is oriented in one direction at one position and in opposite direction in another direction. This kind of arrangement leads to a dipole that stabilizes the crystal lattice of muga wild silk fibre.

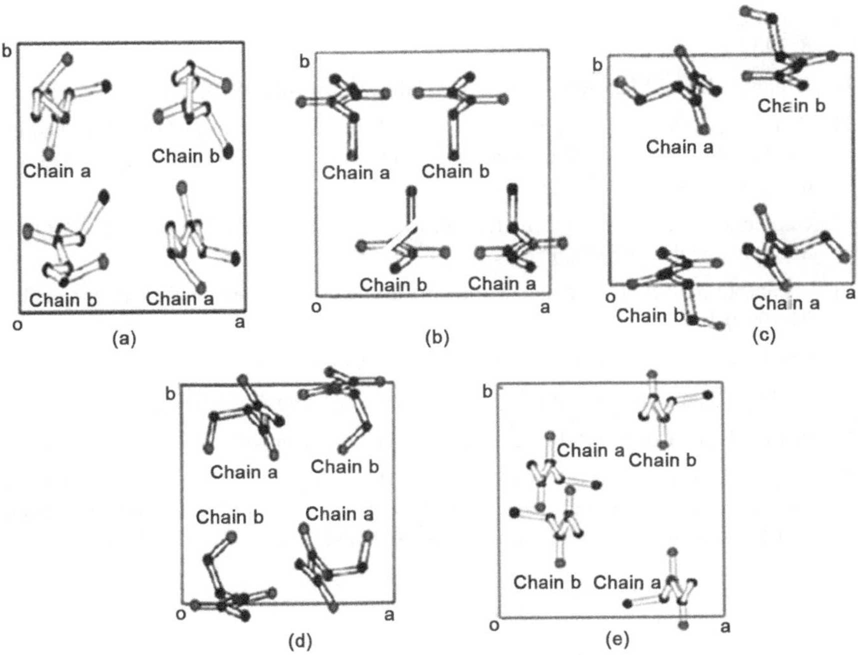

Figure 28.2 Crystal structure (projected along fibre axis) of (a) Muga wild silk fibre, (b) As per Marsh's study, (c) Pure Mysore silk fibre, (d) Bivoltine silk fibre, (e) Tassar wild silk fibre
Light gray – oxygen atom; Dark gray – Nitrogen atom; and (c) carbon atom [17]

References

1. Reddy Narendra and Yang Yiki. Journal of Material Science, 45(2010) 4414.
2. Samitha Maitty, Sagar I Goel, Sobhan Roy, Suvankar Ghorai, Swathi Bhattacharya, Aravind Venugopalan and Anantha K Ghosh. Comparative and Functional Genomics, 2010(2010) Article Id 246738.
3. Singh GP, Zeya SB, Srivastava AK, Prakash B, Ojha NG and Suryanarayana N. Caspian Journal of Environmental Science, 6(2) (2008) 161.
4. Marsh RE, Corey RB and Pauling L. Biochem Biophys Acra, 16(1955)1.
5. Lucas F and Rudall KM. In: Comprehensive Biochemistry, edited by M Florkin & EH Stotz (Elsevier, Amsterdam), 1968, 475.

6. Okuyama K, Somasekhar R, Keiichi Nogochi and Syuji Lchimura. Biopolymers, 59(2001) 310.
7. Sangappa, Mahesh SS, and Somashekar R. Journal of Bioscience, 30(2)(2005)259.
8. Mahesh SS and Somashekar R. Indian Journal of Fibres and Textile Research, 32(2007)143.
9. Parameswaran P and Somasekhar R. ISRN material science, 2011(2011) Article ID 857432.
10. Ivanova MI and Makowshi L. Acta Crystallography, A54(1998)626.
11. Squire J, Al-khayat H, Armott S, Crawshaw J, Denny R, Dover D, Forsyth T, Andrew He, Knupp C, Mant G, Rajakumar G, Rodman M, Shotton M and Windle A. Fibre diffraction review, 11(2003)7.
12. Okuyama K., Keiichi Noguch, Takashi Miyazawa Tashifumi Yui and Kogo Ogawa. Macromolecules, 30(1997) 5849.
13. Takahashi Y, Gehoh M and Yuzuriha K. International Journal of Macromolecules, 24(1999) 127.
14. Smith PJ and Arnott S. Acra Crystallography, A34(1978)3.
15. Pauling L and Corey RB. Proceedings National Academy of Science, 37(1951g) 282.
16. Pauling L and Corey RB. Proceedings National Academy of Science, 39(1953a) 253.
17. Theja Urs G, Ananda HT, Nanda Prakash MB, Byrappa K and Somashekar R. Crystal and molecular structure of muga wild silk fibres based on $[Ala\text{-}Gly]_n$ sequence using LALS technique, Indian Journal of Fibres and Textile Research, 40(2015) 131.

29

Development of silk fabric patterns through new eco-printing method

Summary : There are two carbonized microstructures that result from laser eco-printing technology for silk patterns (SLEP). These are bar-shaped clots and sludge materials with small holes. The bar-shaped clots are produced by the initial melting of raw silk on the silk fabric surface, and the sludge materials with small holes are the combined result of the development of the bar-shaped clot along the vertical and horizontal directions and integration during in-depth printing. Studies have been done on the TG/pyrolysis properties of silk fabric under different atmospheres with regard to the chemical composition and structure of the raw silk. Above a temperature of 280°C, silk reaches a stage wherein there is a rapid weight loss as seen from the thermo-gravimetric curves (TG). The composition and structure of raw silk determines the critical point. The duration of heating and oxygen content greatly influence the pyrolysis speed and degree of weight loss. Hence by setting reasonable printing parameters the printing effects can be obtained. Silk fabric patterns printed by SLEP exhibit yellow chromatically with 10% lightness, and their boundaries are clear and distinct. Carbonization basically takes place on the silk fabric surface. Irrespective of their chemical composition or higher order structures, the fibrous proteins of the irradiated silk fabric melt immediately and carbonize at a high temperature because of the high density of the laser energy used. There are differences in the degree of carbonization of fibrous proteins under different SLEP parameters. But, such differences are not essential ones and occur on the silk fabric surface and hence good printing effects can still be achieved. The technology is prospective and holds promise for development into a new method for forming silk proteins.

29.1 Introduction

Silk considered as the queen of textile fibres faces competition from manmade, synthetic, and other functional fibres. But, due to their special properties like elegance and outstanding texture, silk fabrics are still widely favored by consumers [1–6]. Many silk fabrics have interesting or colorful designs that make them beautiful and lively. To date, such patterns can be created using methods such as hand painting, color printing, spraying, weaving, or embroidery [7–13]. Each of these methods has its own advantages. For example, hand painting can produce unique, colorful unique, colorful patterns, whereas weaving can produce highly reproducible patterns at a high production speed. However, the common feature of each method is the

DOI: 10.1201/9781003735427-29

need for dye. All of these methods require the dyeing of raw silk or fabrics or the spraying of dye onto silk fabrics. Therefore, the use of dye not only makes fibre production processes more complex and costly but also makes these processes harmful to the environment and even to human health [14, 15]. Recently, the researchers proposed an innovative concept involving heat-induced eco-printing (HIEP) or ordinary paper without the use of toner or ink [16–18]. This technology uses the yellowing discoloration of plant fibres and eliminates the environmental pollution caused by the ink used in the printing industry [19]. By testing and analyzing the pyrolysis volatiles of printing paper, they proved that the volatiles produce after HIEP did not include any carcinogens, and hence, HIEP was found to be an environment-friendly technology [20–22].

The chapter highlights the development of a new silk laser eco-printing (SLEP) technology based on heat-induced inkless eco-printing. The microstructure arising from SLEP has been well explored, and the TG/pyrolysis properties, yellowing discoloration mechanism, and also printing effects studied.

29.2 Technical details

Plain crepe satin fabric made from silk has been used for study. A laser has been used for performing laser ablation in order to print silk fabric patterns and measure chromacity. Three different printing speeds have been selected at two different power levels. A color luminance meter has been used to lightness and chromacity at two points on a color block containing the silk fabric sample. An ultra plus field emission scanning electron microscope has been used to observe the microstructure of the color blocks. A special type of calorimeter using nitrogen and air determines the thermogravimetric and differential thermal analysis.

29.3 Microstructure type and formation process

The microstructure of the color blocks is depicted in Figure 29.1. Referring to the figure, the laser power required for producing the materials on the left is lower in comparison with that on the right side. When seen from top to bottom, the laser power and print speed increase.

The SEM images of materials produced without induced printing are shown in Figures 29.2(a) and (b). The partially magnified images of a typical carbonized microstructure is depicted if Figures 29.2(c) and (d). It can be clearly observed that the carbonized microstructures after SLEP are same as

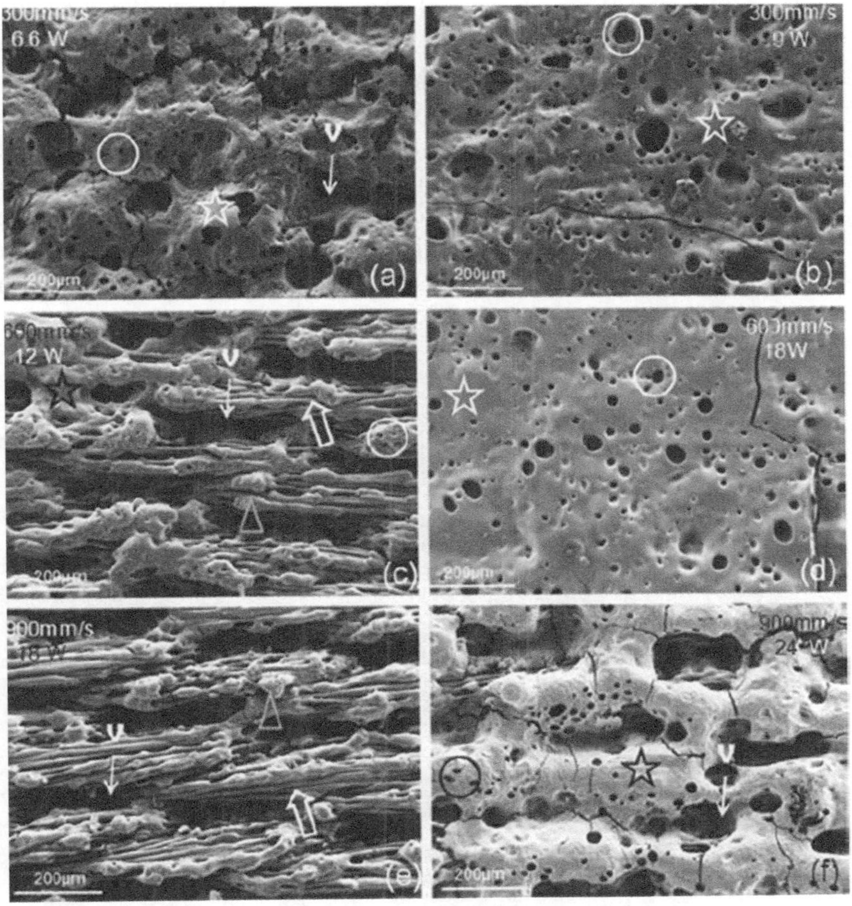

Figure 29.1 Silk fabric showing microstructures having various print speeds and laser power settings (after SLEP) (a–f) [34]
V – interspaces; arrows – fibre; circle – small holes; triangle – clot material and star – sludge

those of sludge materials (indicated by star) as depicted in Figure 29.1. The triangle in the photograph represents bar-shaped clots and the circle represents the many small holes. Figures 29.1(a), (c) and (e) show that while printing at power, more intact fibres remain, as can also be seen (indicated by wide arrow) for the two samples printed at higher speeds [Figures 29.1(c) and (e)]. The finding shows that at high printing speed short duration of heating the samples do not get easily carbonized. As can be seen from Figures 29.1(a), (c) and (e), the sludge formation having many clots and interweaving of the fibres can be observed. Figures 29.1(b), (d) and (f) depicted on the right side the formation of sludge materials, which are essentially integrated together to form sludge materials having many small holes. This finding shows that at the same speed fibre carbonization takes place to a greater extent and results from higher power, and also results in more formation of sludge material. As depicted by Figures 29.1(b), (d) and (f), the laser power has a greater influence on the extent of carbonization as compared to the printing speed. Whereas, as depicted in Figures 29.1(c) and (e) in the case of samples treated with lower laser power the clots are primarily distributed over them. The laser printing does not create the cracks in sludge materials, but are created during preservation process by exogenic action.

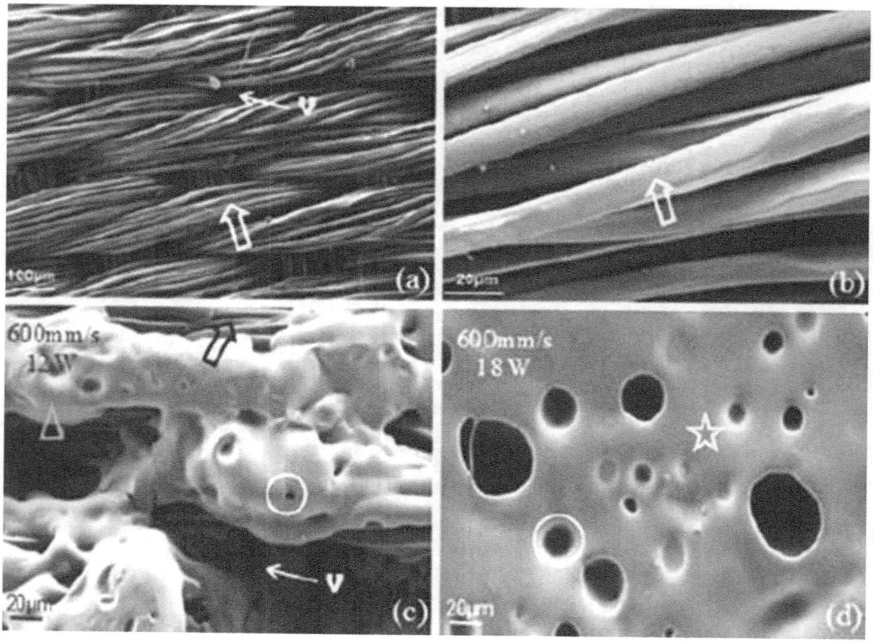

Figure 29.2 Structure of silk fabric before laser printing (a,b) [34], (b) magnified image, and (cd) a typical carbonized microstructure after SLEP at the laser power settings (marks similar to Figure 29.1)

On observing Figures 29.2(a), (c) and (e), the interspaces among the clots have not been created by carbonization decomposition but by the weaving of warp and weft yarn, which are depicted in Figures 29.2(a) and (b) as structural images of the unprocessed silk. Hence, the carbonization of silk fabric subjected to laser irradiation is supposed to take place in the following manner:

(a) The raw silk on the surface of silk fabric starts to soften and bulges to form clots

(b) With the increase in laser power or duration of print, degradation takes place along the vertical and horizontal directions till the sample ultimately becomes an even sludge material having small holes.

29.4 Factors affecting thermogravimetric/pyrolysis properties of silk fabric

The TG and DTA curves of the silk fabric derived from nitrogen and air have been depicted in Figures 29.3a and b. The TG curves of the critical temperature, speed and rate of weight loss comprise of five stages, even though they vary between the two different environments. Their temperature and weight loss at each critical point have been determined. During the first stage, weight loss takes place slowly with a ratio of 7% because of water evaporation [23, 24]. From the second to fifth stage, having a critical point of −280°C, weight loss takes place rapidly because of the pyrolysis of raw silk, as depicted in Figures 29.3(a) [25, 26]. This result is consistent with the TG results analyzed for raw silk in nitrogen [27, 28]. This consistency suggests that the critical points in the TG curves are dictated by the composition and structure of raw silk. Raw mulberry silk consists of 18 amino acids which basically are glycine (43%), alanine (32%), serine (15%), and tyrosine (12%) [29, 30]. The other contents exist less than 2%, except valine (3%), and half of the other amino acids are present in quantities of less than 1%. Amino acids are themselves crystals having melting points ranging from 200°C to 300°C. Raw silk is a fibrous protein which contains 18 amino acids that form a peptide chain (H chain) having a high molecular weight and 1–3 peptide chains having a low molecular weight. The H chain has the appearance of a β-sheet or α-helical structure. Each amino acid in raw silk is integrated into a chain formed by large molecules whose melting point is ~300°C. But, Figure 29.3(a) depicts the second to fifth stage involving the phenomenon of rapid weight loss above ~280°C. During the second to fifth stages, the various rates of weight loss noticed are mainly determined by the structural forms of raw silk. There are three distinct regions in raw silk.

These are crystallization, non-crystallization and transition regions [31]. Glycine, alanine and serine are present in the crystallization region in the ratio 3:2:1, with small side chains. Amino acids having large side chains or polar radicals are present in the non-crystallization region. Raw silk has a crystallinity of 40–50%, and hence its pyrolysis can take place during the second stage in the non-crystallization region. Depending on the heating atmosphere used, pyrolysis during crystallization is different from that in the transition regions. Even at temperatures up to 800°C, a pyrolysis residue content above 32% is present in the final material, in nitrogen. But heating below 650°C in air, the final material exhibits a residue content of less than 5%. This phenomenon occurs because, as compared to a nitrogen atmosphere, oxidation or hydrolysis accompanies pyrolysis in air, which accelerates the speed at which the critical temperature is reached as well as the rate of weight loss (Figure 29.3). Hence, temperature coupled with high oxygen content highly accelerates the speed of pyrolysis and the weight loss of raw silk.

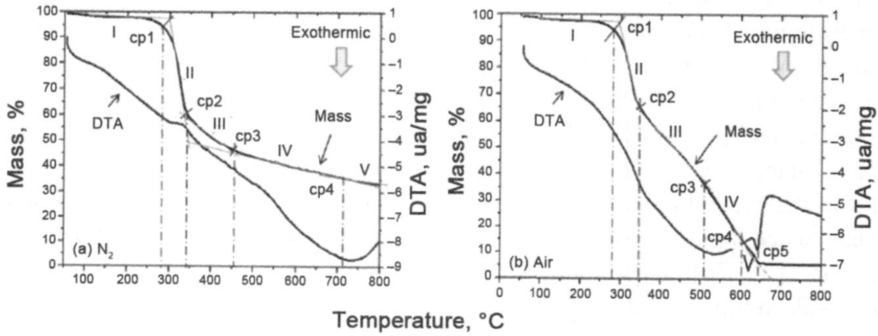

Figure 29.3 Thermogravimetric and differential thermal analysis of silk fabric (a) Nitrogen and (b) Air

29.5 Printing effects, yellowing and discoloration mechanism

Figure 29.4(a) shows the lightness L* of each sample together with the color blocks. Figure 29.4(b) shows several printed samples. Figure 29.5 shows the carbonized microstructure of raw silk near the print boundary. Figure 29.4(a) shows that the L* values ranged from 10 to 35 under the experimental conditions chromaticity results show that the values of x range from 0.42 to 0.44, whereas the values of y range from 0.38 to 0.40 and are primarily distributed in the range of yellow chromaticity without any large fluctuations. Differences in the color blocks processed at different powers are primarily due to different lightness L* values [Figure 29.4(a)]. The carbonized microstructure along the print boundary shows several swollen fibres (Figure 29.5), and the transition region from the non-printed to the normally carbonized material is extremely narrow. This effect is due to the small diameter of the laser spot used and the concentration of the laser energy at the printed points. Thus, good printing effects are obtained under these experimental conditions despite the disparity in chromacity and lightness values (Figure 29.4) between the print samples.

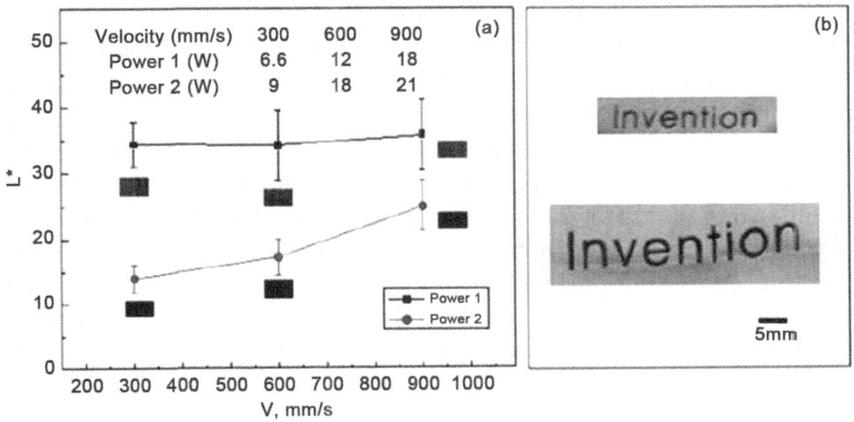

Figure 29.4 Laser ablation conditions and printing effects [34]
(a) Color blocks and their lightness L* values, and (b) Printed samples

As discussed above, the degree of weight loss and pyrolysis (change in composition and structure) of silk fabric vary due to different thermal effects and the atmospheric conditions (Figure 29.3). This result is due to the use of a laser with a thin light beam and concentrated energy and because the printing time is extremely short (the duration of laser ablation was less than 1 ms for each print point) with SLEP technology. In other words, SLEP has features that include a high concentration of energy and a short heating time. Therefore, although the experiment was performed in air, a condition in which silk generally undergoes oxidation and pyrolysis (pyrolysis), fibrous proteins on the surface of the silk fabric that were irradiated (printed) melted immediately (Figure 29.1) regardless of the composition or higher order structure in the crystallization and the non-crystallization regions due to the high laser energy density.

Even peptide chains, composed of large molecules in raw silk fibres formed by amino acids and all types of polar radicals in amino acids, are directly destroyed. Moreover, as expected, protein fibres are carbonized at high temperatures. Hence, although there are differences in both the parameters of the SLEP process and the microstructure of the silk fabric after printing (Figure 29.1), the differences primarily concern the quantity (area and depth) of the carbonized protein fibres and are not essentially distinct, because the laser parameters used in this study are determined based on preliminary experiments. Thus, from a macroscopic view, good printing effects are obtained even though there are differences in lightness and chromaticity. For example, the lightness values are approximately 10 under the three conditions involving high power. The lightness value is near the 0 value of black, as high as 90%. In addition, analysis of the microstructures of the silk fabric before and after printing (qualitative analysis) indicates that carbonization occurs primarily on the surfaces of silk fabric (warp or weft) and that the carbonization depth is approximately 1/10–1/4 of the fabric thickness (Figure 29.5, estimated by visual inspection). Therefore, silk fabric after SLEP should be sufficiently strong for use. This result is consistent with a visual assessment of the print samples. Furthermore, the advantages of using SLEP are not only the simplicity of the production processes but also that it does not require dye materials. Furthermore, as a consequence, environmental pollution and the harm to psychological health from its use are eliminated. Thus, SLEP technology could become a new method for forming silk fabric patterns.

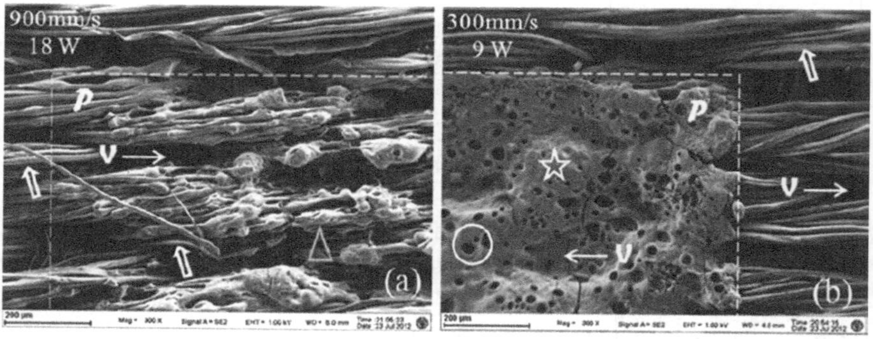

Figure 29.5 Carbonized silk fabric showing microstructure (boundary lettered) after SLEP [34] (a) Power 1, and (b) Power 2
(white dotted line divides the print lines; the letter P denotes the printed side; other marks similar to Figure 29.1)

Furthermore, yellowing is caused by the radical oxidation of amino acids with aromatic nuclei in fibroin or by the degradation of fibroin peptide chains. Amino acids with aromatic nuclei (yellow) are found in the non-crystallization region. Amino acids that undergo molecular chain breaking are primarily those that occupy a large volume, such as tyrosine, valine, praline, glutamic acid and threonine [32]. As discussed above, laser printing processes induce unique thermal behaviors [33]. It is assumed that the molecular structures in the proteins are directly damaged. However, because the process is related to the mechanism of molecular pyrolysis, for example, in terms of the strength of thermal action during the printing process, the decomposition process of fibres in silk fabric and its environmental protection property affect the formation of the final materials as well as yellowing and discoloration. The optimization of SLEP parameters should be the subject of further in-depth research.

References

1. Kothari VK, Sengupta AK, Srinivasan J and Goswami BC. Textile Research Journal, 59(1989) 292.
2. Tsuchiya I. Matsumosa Y, Tonumi K and Harakawa K, Textile Research Journal, 61(1991) 131.
3. Ke GZ, Xu WL and Yu WD. Indian Journal of Fibres and Textile Research, 33(2008) 185.
4. El Molla MM, El Khatib EM, El Gammal MS and Abdel Fattah SH. Indian Journal of Fibre and Textile Research, 36(2011) 266.
5. Venkataraman K, Govindarajan S, Lingasamy K, Balakrishnan M, Tirumanasekaran D and Ramasamy A. Indian Journal of Fibre and Textile Research, 39(2014) 172.
6. Das D, Bhattacharya S, Chandra M, Sankar R. Indian Journal of Fibre and Textile Research, 31(2006) 559.
8. Chataw U. International conference on digital printing technology, (2002) 125.
9. Ku F. Silk, 49(2012) 59.
10. Wu HH. Silk, 49(2012) 37.
11. Quian XP. Song Brocade Yun Brocade Silk 48(2011) 1.
12. Liu AD, Li B and Qiu YP. Silk, 49(2012) 50.
13. Li D, Zhang JQ and Li XJ. Silk, 49(2012) (50).
14. Zhou Q and Zhao YC. Journal of Environmental Health, 22(2005) 229.
15. Cucci I, Baschi A, Arosta C, Berlini F, Freddi G and Catellani M. Synthesis of mioletals, 159(2009) 246.
16. Chen JX, Wang Y, Xie J, Meng C, Wu G and Zu Q. Carbohydrate polymer, 89(2012) 849.
17. Xie J, Chen JX, Wang Y, Liu YF, Noori MN and Pan L. Cellulose chemical technology, 48(2014) 577.
18. Chen JX, Xu LN, Xie J, Wang Y, Zu Q and Meng. Cellulose chemical technology.
19. Chandgnya MH, Angersa DA and Beuchamp CJ. Soil biology and biochemistry, 32(2000) 1561.
20. Pan L, Chen JX, Wan CF, Ren H, Zhal HM and Wang Y. Cellulose chemical technology.
21. Chen JX, Xie J, Wu G, Ren H and Wang Y. Cellulose 21(2014) 2871.
22. Chen JX, Xie J, Pan L, Wang X, Xu LN and Lu Y. Journal of Wood Chemical Technology, 34(2014) 202.

23. Wu CY, Feng ZZ, Wan SM, and Wu JY. Silk chemistry, 2nd edn. (China textile press Beijing), 1990, 6–90.

24. Zuo BQ, Zha M and Yang YM. Silk, 40(2003) 35.

25. Sashina ES, Janowska G, Zaborski M and Vnuchkin AV. Journal of Thermal Analysis and Calorimetry, 89(2007) 887.

26. Wu SQ, Li MY, Fang BS and Tong H. Carbohydrate polymers, 88(2012) 496.

27. Zhang H, Magoshi J, Becker M, Chen JY and Matsunaga R. Journal of Applied Polymer Science, 86(2002) 1817.

28. Freddi G, Tsukada M and Beretta S. Journal of Applied Polymer Science, 71(1999) 1563.

29. Liu GF, Wang XL and Hu C. Journal of Zhejiang Sichou Institute of Technology, 10(1993) 1.

30. Luo WZ and Feng YF. Journal of Zhejiang Sichou Institute of Technology, 7(1990) 52.

31. Wray LS, Hu X, Gallego J, Georgakoudi I, Omenetto FG, Schmeditt D and Kaplan DL, Journal of Biomaterial Research B, 99(2011) 89.

32. Reddy N and Yang YQ. Journal of Material Science, 45(2010) 6617.

33. Chen JB and Peng RL. The Theory and Application of Laser (Electronic industry press, Beijing) 2010,163.

34. Jinxiang C, Chuang M, Juan X, Le P, Dong Z and Junnan C. Laser eco-printing technology for silk fabric patterns, Indian Journal of Fibres and Textile Research, Volume 41, March 2016, 78–83.

30

Effect of some key factors on scouring of sericin in muga silk cocoons

Summary : Comparison has been made on the degumming of muga silk cocoons under high temperature and pressure conditions and sodium carbonate. The former reveals a better degumming loss (23.67%), than the latter (22.28%). After degumming in autoclave, the quality of fibre remains strong with improved lustre, which is typical characteristic of muga silk. Hence superior quality of muga silk has been produced in autoclave degumming with regard to sericin loss, surface smoothness and fibre strength compared to contrast to conventional degumming. Hence the findings recommend the use of autoclave cocoon degumming on industrial scale basis as the process is cost effective and chemical free leaving no hazardous effect on the fibre as well as on the environment.

30.1 Introduction

There are two types of structural proteins in the silk cocoon. One is the inner core crystalline nature fibroin and the other is the outer amorphous sericin which is sticky. The fibroin chains get oriented by the elongational flow of spinning, and the fibroin in liquid state is converted into partly crystalline, insoluble fibrous filaments of solid form [1]. The sericin protects the fibroin core and helps to reduce the shear stress and also absorbs the squeezed water from the stretched fibroin in the process of fibre creation [2]. The whole processing of silk from cocoons to the finished product entails the following steps of reeling, weaving, degumming, dyeing and finishing. It is necessary to remove sericin from the fibroin in order to achieve a superior quality of silk with lustre and elegant drape. Degumming is basically a thermomechanical process to carry out scouring of sericin. The protein of sericin gets hydrolyzed into its subsequent amino acids and is solubilized in the degumming medium during the degumming process [3, 4]. Sodium carbonate is used as degumming substance in industries. Many acidic, alkaline and neutral proteases have been used as degumming agents on silk fabric over the years. Proteases of alkaline nature are found to be more advantageous than those of acidic and neutral nature for achieving uniform degumming and good silk quality [5, 6]. However, high cost and low performance in silk handling results in limited use of enzymes in the industrial scale. The consumption of chemicals by most of the previously mentioned methods causes serious pollution to the

DOI: 10.1201/9781003735427-30

land and water systems and thereby affects the environment [7]. Indigenous muga silk created by the sericogenous insect *Antheraea assamensis* Helfer is found more in the North Eastern part of India. Muga silk possesses some striking features like attractive golden color, high mechanical properties and long durability. Till date degumming of muga silk is reported to be carried out by using sodium carbonate and a biosurfactant (reetha) as the degumming agent [8, 9]. The present study discusses the effect of degumming muga silk cocoon under autoclave conditions of high temperature and high pressure. There are no toxic harmful effects on fibre and environment when degumming under high temperature and pressure conditions and thus proves economical; whereas, in the case of chemical degumming this is not so.

30.2 Technical details

Freshly spun cocoons of *Antheraea assamensis* have been used. The degumming has been carried out using sodium carbonate solution adopting the normal procedure. Autoclave method has been used for degumming under prescribed conditions [26]. The silk cocoons have been reeled in wet condition at room temperature [26]. The reeled silk length and the number of breaks during reeling have been studied. The following studies have been conducted:

(a) Tensile properties
(b) FTIR spectroscopy
(c) Thermogravimetric analysis
(d) Differential scanning calorimetry
(e) Surface morphology

One way ANOVA has been used to statistically analyze the data obtained from studies. Differences of less than 5% between experimental groups have been considered statistically significant and less than 0.1% difference as highly significant.

30.3 Effect of degumming

The efficiency of degumming quality is determined by degumming loss % degumming loss is higher in the case autoclave degumming (23.67) than for sodium carbonate degumming (22.28). The one-way ANOVA analysis shows that the results highly significant at p value <0.001. The sericin of muga silk fibre due to its random structures, can be completely hydrolyzed into its subsequent amino acids under high temperature and pressure without further deterioration to the fibre content [10, 11]. Silk fibre becomes mechanically weak by sodium carbonate degumming which affects the non-covalent bonds of silk fibroin [26]. Degumming of silk under high temperature and pressure or boiling in sodium carbonate solution results in yield of continuous fibre by effective removal of the adhesive gum from the silk fibre. The total length of reeled silk is obtained to be 280 m in autoclave degumming and 235 m in sodium carbonate degumming. During reeling the number of filament breakages is found to be more (30) in case of sodium carbonate degumming than that of autoclave degumming (10 only). The breakage of fibre arises due to sodium carbonate attacking the non-covalent bonds of silk fibre, and consequently affects its mechanical strength and fibre thickness [13].

30.4 Mechanical properties

In the case of sodium carbonate degummed fibre, the tensile strength, elongation, Young's modulus and toughness are found to be 4.01 ± 0.22 g/den, 30.85 ± 0.47%, 70.39 ± 2.71 and 0.719 ± 0.18 g/den, respectively. In comparison with conventional degumming, the degumming under high temperature and pressure conditions certainly augmented the tensile behavior of the silk cocoon fibre with tensile strength 4.80 ± 0.30 g/den, elongation 33.53 ± 0.52%, Young's modulus 85.30 ± 3.09 g/den and toughness 0.994 ± 0.56 g/den. Partial damage of cross-sectional area and microstructure of core fibroin occurs due to reduction in tensile strength of fibre after sodium carbonate treatment [14]. Autoclave degumming sequestered sericin protein from silk cocoon fibres without disturbing the chemical bonds and hence the mechanical strength of fibroin filaments. Figure 30.1(a) depicts the FTIR spectrographs of degummed muga silk fibres.

The helix/random coil configuration of silk protein is responsible for the amide I peak of sodium carbonate and autoclave treated muga fibre at 1657 cm^{-1} and 1647 cm^{-1}, respectively [15, 16]. The C=O stretching vibration (about 80%) causes the amide I vibration, and there is small influence from the out of phase CN stretching vibration, the CCN deformation and the NH in-plane bend. The amide II band observed at 1525 cm^{-1} and 1515 cm^{-1} for sodium carbonate and autoclave degummed fibre, respectively, is significant of β-sheet configuration structure [18]. Similarly, amide III mode is the inter-phase combination of the NH bending and the CN stretching vibration with minimum amount of contribution from the CO in-plane bending and the CC stretching vibration [19]. It is possible to confirm the amide III band at 1233 cm^{-1} for both degummed fibres because of oriented β-sheet structure of silk fibre. Thus it can be observed that there is no change in the intrinsic molecular conformation of muga cocoons caused by degumming with both the methods mentioned.

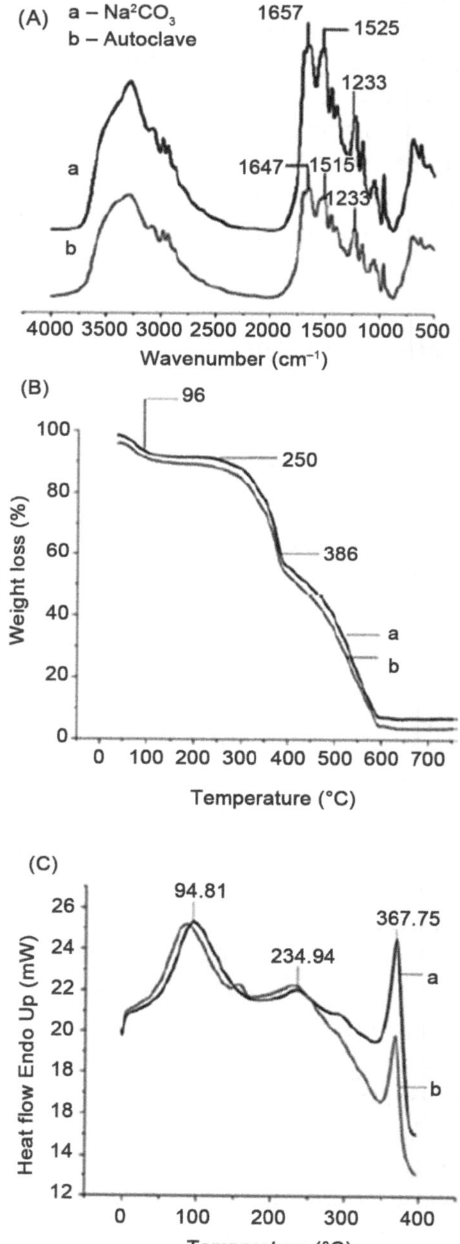

Figure 30.1 (A) FTIR spectroscopy, (B) Thermogravimetric analysis, and (C) Differential scanning calorimetry of degummed muga silk fibres [26]

30.5 Thermogravimetric studies

The TGA analysis reveals that because of evaporation of moisture there is initial mass loss step in the fibres at around 96°C [Figure 30.1(b)]. This is followed by the weight loss occurring between 250°C and 386°C. This is evident owing to the final degradation of highly oriented β-sheet crystalline structure of silk [20].

30.6 Differential scanning calorimetry studies

As can be seen from Figure 30.1(c), at about 94°C the first endothermic peak is observed during the DSC investigation, which confirms the loss of water bound to the amorphous region of silk through hydrogen bonding. The major endothermic peaks at 234°C and 366°C above the glass transition temperature T_g (190–200°C) signifies phase transition from amorphous to crystalline one and finally complete degradation of stable β-sheet orientation [21–24]. When considering the thermal behavior of silk fibres, the difference is not so significant.

30.7 SEM studies

Figure 30.2 depicts the surface micrographs of the degummed muga silk cocoon fibres. There is removal of sericin as seen from the surface architecture of the sodium carbonate degummed fibre, while still leaving certain amount of sericin patches on its surface topology. But, there is perfect degumming wherein sericin molecules break away from the fibroin filaments as revealed by the topography of autoclave degummed fibre. Degumming by autoclave shows that the fibre surface is very smooth, uniform and clean without damage to the silk fibroin, exhibiting only longitudinal striations characteristic of fibrillar structure [25].

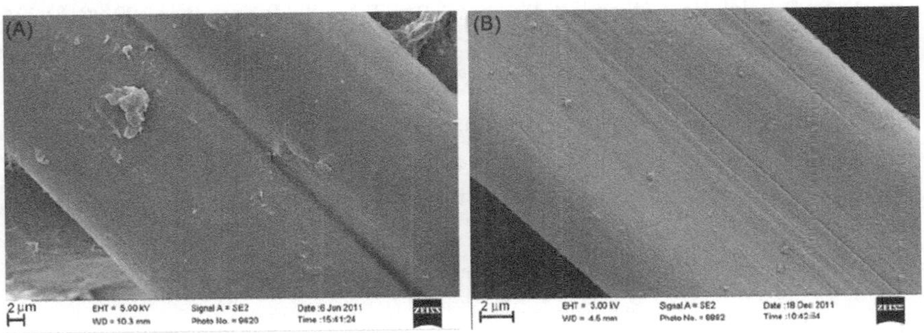

Figure 30.2 SEM micrographs of (A) sodium carbonate and (B) autoclave degummed muga silk fibre [26]

References

1. Ikada Y and Tsuji H. Macromolecule rapid communication, 21(2000) 117.
2. Jin HJ and Kaplan DL. Nature, 424(2003) 1057.
3. Chopra S and Gulrajani ML. Indian Journal of Fibres and Textile Research, 19(1994) 76.
4. Duran N and Duran M. Reviews in progressive color related topics, 30(2000) 41.
5. Gulrajani ML, Gupta SV, Gupta A, and Suri M. Indian Journal of Fibres and Textile Research, 21(1996) 52.
6. Gulrajani ML, Sen S, Sona A and Suri M. Indian Journal of Fibres and Textile Research, 23(1998) 52.
7. Mahmoodi NM, Arami M, Mazaheri F and Rahimi S. Journal of Cleaner Production, 18(2010) 146.
8. Devi D, Sarma NS, Talukdar B, Chetri P, Baruah KC and Das NN. Journal of Textile Institute, 102(2011) 527.
9. Sarma MB, Gogoi SB, Devi D and Goswami B. Journal of Scientific and Industrial Research, 71(2012) 270.
10. Tanaka T. Journal of Thermal Analysis and Calorimetry, 70(2002) 825.
11. Xiaomei Z, Berghie I and Wyeth P. Journal of Cultural Heritage, 12(2011) 408.
12. Yamada H, Nakob H, Takasura Y, and Tsubouch K. Material Science Engineering, 14(2001) 41.
13. Jiang P, Liu H, and Wang C. Material Letters, 60(2006) 919.
14. Khan MR. Bioresource Technology, 101(2010) 8439.
15. Trettinikov ON and Tamada Y. Langmiur, 17(2001) 7406.
16. Teramoto H, and Miyazawa, Biomacromolecules, 6(2005) 2049.
17. Krimm S and Bandekar J. Advanced Protein Chemistry, 38(1986) 181.
18. Freddi G, Monti P, Nagura M, Gotoh Y, and Tsukada M. Journal of Polymer Science, Part B: Polymer Physics, 35(1997) 841.
19. Barth A. Biochemistry Biophysics Acta, 1767(2007) 1073.
20. Dutta S, Talukdar B, Bharali R, Rajkhowa R and Devi D. Biopolymers, 99(2013) 326.
21. Tsukada M, Gotoh Y, Freddi G, Matsumura M, Shiozaki H and Ishikawa H. Journal of Applied Polymer Science, 44(1992) 2203.
22. Kweon H, Um IC, and Park YH, Polymer, 41(2000) 7361.
23. Tsukada M. Journal of Polymer Science Part B: Polymer Physics, 26(1988) 949.
24. Lu Q, Hu X, Wang X, Kluge JA, Lu S, Cebe P and Kaplan DL. Acta Biomaterials, 6(2010)1380.
25. Arai T, Freddi G and Innocenti R. Journal of Applied Polymer Science, 91(2004) 2383.
26. Choudhary M and Dipali D. Impact of high temperature and pressure on sericin scouring of muga silk cocoons, Indian Journal of Fibres and Textile Research, Volume 41, March 2016, 93.

Index

E

F

G

H

I

K

L

M

T

U

V

W

X

Y

For Product Safety Concerns and Information please contact our EU representative GPSR@taylorandfrancis.com Taylor & Francis Verlag GmbH, Kaufingerstraße 24, 80331 München, Germany